MW01503780

Searching for Habitable Worlds

An Introduction

Searching for Habitable Worlds

An Introduction

Abel Méndez
Planetary Habitability Laboratory,
University of Puerto Rico at Arecibo, Puerto Rico

Wilson González-Espada
Department of Mathematics and Physics,
Morehead State University, Morehead, KY, USA

Morgan & Claypool Publishers

Copyright © 2016 Morgan & Claypool Publishers

All rights reserved. No part of this publication may be reproduced, stored in a retrieval system or transmitted in any form or by any means, electronic, mechanical, photocopying, recording or otherwise, without the prior permission of the publisher, or as expressly permitted by law or under terms agreed with the appropriate rights organization. Multiple copying is permitted in accordance with the terms of licences issued by the Copyright Licensing Agency, the Copyright Clearance Centre and other reproduction rights organisations.

Rights & Permissions
To obtain permission to re-use copyrighted material from Morgan & Claypool Publishers, please contact info@morganclaypool.com.

ISBN   978-1-6817-4401-8 (ebook)
ISBN   9781643278957 (print)
ISBN   978-1-6817-4403-2 (mobi)

DOI   10.1088/978-1-6817-4401-8

Version: 20160401

IOP Concise Physics
ISSN 2053-2571 (online)
ISSN 2054-7307 (print)

A Morgan & Claypool publication as part of IOP Concise Physics
Published by Morgan & Claypool Publishers, 40 Oak Drive, San Rafael, CA, 94903, USA

IOP Publishing, Temple Circus, Temple Way, Bristol BS1 6HG, UK

To all explorers, we are just a dust particle in the universe, make it count.

Contents

Preface   ix

Acknowledgments   x

About the authors   xi

1   Exploring Earth and beyond   **1-1**

1.1   Curiosity   1-1

1.2   Observations from afar   1-3

1.3   Personal visits   1-6

1.4   Robotic visits   1-8

     References   1-10

2   The ABC of exoplanets   **2-1**

2.1   What is a planet?   2-1

2.2   What is an exoplanet?   2-3

2.3   Detecting exoplanets   2-5

     References   2-10

3   When is a planet habitable?   **3-1**

3.1   What is life?   3-1

3.2   The concept of habitability   3-4

3.3   Measuring habitability   3-9

     References   3-12

4   Cataloguing habitable exoplanets   **4-1**

4.1   The first habitable worlds?   4-1

4.2   The habitable exoplanets catalog   4-2

4.3   The periodic table of exoplanets   4-4

     References   4-6

5   Potentially habitable worlds   **5-1**

5.1   The planets we know by mass   5-1

5.2   The planets we know by size   5-8

     References   5-21

6   And the search goes on…   **6-1**

6.1   The observatories   6-1
6.2   NASA's Kepler   6-1
6.3   NASA's TESS   6-3
6.4   ESA's Plato   6-4
6.5   NASA JWST   6-5
      Reference   6-6

Preface

The hunger for exploring the unknown and answering nature's toughest questions is one of humanity's greatest assets. This process has led to theoretical, experimental and practical discoveries in all areas of science, discoveries that have exponentially expanded our limited human senses.

In astronomy, one of the quests that has tested our technological abilities to the limit is the detection and characterization of exoplanets, planets that orbit stars other than our Sun. This is so because the distances between our Solar System and other similar systems in our own Galaxy are simply stupendous, and to even consider the possibility of exoplanets in other galaxies and galaxy super clusters is mindboggling.

So far, over 2000 exoplanets have been discovered, from Jupiter-like behemoths to rocky bodies not unlike Earth or Mars. Their orbital periods are just as varied, some exoplanets zoom around their orbit in as little as three hours, while others orbit sluggishly, their journey taking centuries for each revolution. Some exoplanets are hotter than Venus or colder than Uranus. Although these celestial bodies are as diverse as snowflakes, astronomers have obtained enough information about their physical parameters to start a classification process, a detailed Periodic Table of Exoplanets.

Intriguingly, a few exoplanets are found at interesting locations within the habitable zone. They are neither scaldingly hot nor frigidly cold; and they are neither too light to lack atmospheres or magnetic fields, nor too heavy that gravity forces become unbearable. These exoplanets could potentially be suitable for life as we know it, or might harbor life forms we have not even conceived of in our wildest dreams. What does it take to consider a planet potentially habitable? If a planet is suitable for life, could life be present? Is life an inevitable result of the laws of science? What are the latest science paradigms and the wildest hypotheses that astronomers are considering regarding potentially habitable worlds?

Searching for Habitable Worlds provides both the general public and astronomy enthusiasts with a richly illustrated synthesis of the most current knowledge regarding the process of finding extrasolar planets, measuring their physical and chemical properties, and evaluating to what extent they might fit habitability parameters. After reading this book, we expect young people to see astronomy as a cool and obtainable career path, full of wonder and ripe for discovery. We expect people to become even more curious about science and how it has changed how we see the natural world.

Most people want to know if we are alone in the Universe. *Searching for Habitable Worlds* does not have a definitive answer, but will show readers the very first places we should look for life, as we know it.

Acknowledgments

We would like to thank Dr Jennifer J Birriel, Professor of Physics from Morehead State University, for editing the manuscript and providing insightful suggestions for improvement.

We thank the Department of Mathematics and Physics of Morehead State University and the Center for Research and Creative Endeavors of the University of Puerto Rico at Arecibo for their professional support and providing a supportive academic environment.

We would also like to show our gratitude to our families who lovingly tolerated our long hours away from home, researching and writing the manuscript for this book.

About the authors

Abel Méndez

Professor Abel Méndez is an Associate Professor of Physics and Director of the Planetary Habitability Laboratory, Department of Physics and Chemistry, University of Puerto Rico at Arecibo, Puerto Rico. He performs research on planetary habitability, exoplanets and astrobiology in general.

Wilson González-Espada

Dr Wilson González-Espada is an Associate Professor of Physics and Science Education, Department of Mathematics and Physics, Morehead State University, Morehead, KY, USA. His scholarly interests include science communication, the public understanding of science and physics education research.

Chapter 1

Exploring Earth and beyond

1.1 Curiosity

Before going into detail about what is currently known about potentially habitable worlds, it is important to reflect on why humans are so interested in them in particular, and in space more broadly. In one word, the answer is curiosity.

Curiosity is defined as a compulsion to understand our surroundings, a strong eagerness to learn or understand something, or a quest for knowledge that is driven by the novelty, complexity, comprehensibility and unexpected nature of 'interesting' external stimuli [1, 2]. Curiosity, seen by some as the drive to move away from a state of ignorance [3], includes three main types of behaviors: epistemic observation (observing, experimenting), consultation (asking others, seeking sources of information) and direct thinking (prolonged thinking on a single subject) [4].

Many organisms have an exploratory curiosity [5]. But most would argue that humans are the only species that can harness the products of curiosity, what English philosopher Thomas Hobbes described as 'the continual and indefatigable generation of knowledge' [1], and reinvest them to produce exponential amounts of more curiosity (see figure 1.1).

There are many examples of the exponential growth of knowledge that is fed through curiosity. One of them was the development of animal collections and plant specimens. What started as 'cabinets of curiosities' in the 17th century grew to become enormous private and public science museums world-wide. Another one was the compilation of scientific information in written form, starting from the science treatises of Ptolemy, Anaximander and Aristotle, and culminating with the terabytes of scientific information stored on the internet and in 'the cloud'. How about the growth and complexity of technology? Ancient Greeks who marveled at the complexity of the Antikythera mechanism (figure 1.2) would be mesmerized by today's manufacturing, communications, medical and space technologies.

When talking about space exploration, many scholars have pointed out that the main motivations to engage in it are war, greed, or the celebration of power [6].

1-1 © Morgan & Claypool Publishers 2016

Figure 1.1. Children are born scientists; a child's curiosity is unbridled by preconceptions or expectations. Photo courtesy of Mads Bodker.

Figure 1.2. The Antikythera mechanism, a 200 BC mechanical computer that used gears to map the positions of some celestial bodies at a given date. Photo courtesy of Tilemahos Efthimiadis, National Archaeological Museum, Athens, Greece.

However, it could be argued that curiosity is, at least, a much more fundamental and benign driver of human exploration. Part of the reason for this curiosity is how difficult it is to explore 'out there'; regions and objects in the Universe with temperatures near absolute zero or millions of kelvins, places so far away that their light has not reached us yet, celestial objects so sneaky that they emit waves we cannot see and even 'stuff' unlike any type of matter we know.

And yet, we have learned a lot about our Universe, near and far. Human curiosity has led not only to the drive to explore space, but to the development of manned and unmanned technologies that are capable of collecting data from the Earth's surface, Earth's orbit, or by visiting faraway places.

1.2 Observations from afar

One of the easiest ways to explore space is to see it with our own eyes. Thousands of years ago, ancient civilizations already knew that there were two main types of small celestial bodies. The first type, stars, always moved east to west in the night sky, almost always kept the same brightness and were always at the same distance from each other; that is, no relative motions were observed. The second type, planets, did show relative motion with respect to other stars and could be brighter or dimmer than normal. Some of them occasionally showed retrograde or backwards motion, at least temporarily.

Eventually, instruments were designed to locate and measure the position and motion of stars and planets (figure 1.3). One of these is called the planispheric astrolabe and it was used to estimate time, predict sunrise and sunset, locate planets and identify star placement and altitude in the sky. Basically, an astrolabe has a map

Figure 1.3. The members of the prime crew of the Apollo 13 lunar landing mission demonstrate the astrolabe (right) and the octant (left), two essential instruments in early visual astronomy. Photo courtesy of NASA.

of the local sky engraved on a flat disk that rotates over another disk with an intricate pattern of arcs that represents local horizontal coordinates. On the reverse side, the instrument provides an altitude meter. This two-in-one instrument has many practical astronomical applications, such as the determination of the time from a single altitude measurement of the Sun or a star [7]. The astrolabe evolved into smaller, simpler instruments such as the octant, or reflective quadrant.

Another instrument of immeasurable importance in visual astronomy is the telescope. Improved and popularized by Galileo Galilei, the refractive telescope uses two lenses to show a magnified image of celestial objects. In 1610, Galileo published the booklet *Sidereus Nuncius*, in which he shared his telescopic discoveries in the sky, including Jupiter's four largest moons, topographic details of our moon and the myriad stars that make up the Milky Way. In the 1660s, Isaac Newton invented the reflecting telescope using mirrors instead of lenses to obtain sharper images. It uses a large mirror to collect and focus the light into a secondary mirror and an eyepiece. An advantage of the reflecting telescope is its portability; it is an essential tool for 'star parties', astronomy outreach to schools and the general public (figure 1.4).

Scientists soon noticed that the larger the area of the lens or mirror, the more light could be focused on an eyepiece, improving the brightness as well as the resolution of the sharper images. Since lenses cannot be made very large due to weight and aberration issues, most large telescopes are of the reflective variety. One of the largest optical telescopes is located at the W M Keck Observatory, on the summit of Hawaii's dormant Mauna Kea volcano. The Keck telescopes' primary mirrors are 10 m in diameter, divided into 36 hexagonal segments that work in concert as a single unit.

Figure 1.4. Children and adults alike enjoy attending 'star parties' like this one, offered in Old San Juan, Puerto Rico. Photo courtesy of Eddie Irizarry/Astronomical Society of the Caribbean.

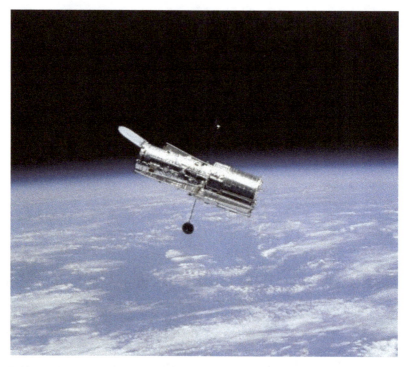

Figure 1.5. The Hubble Space Telescope (HST), after deployment on the second servicing mission (HST SM-02). Note the telescope's open aperture door. Photo courtesy of NASA.

One disadvantage of Earth-based telescopes is that Earth's atmosphere is not perfectly transparent. Turbulence and dust distort astronomical images. One potential solution to this problem is to build telescopes on top of the highest mountains, thereby reducing the amount of atmosphere that light must go through before it arrives at the instruments. Another option is to have no atmospheric interference at all, that is, placing a telescope in orbit around Earth.

The Hubble Space Telescope (figure 1.5), essentially a Newtonian telescope, has been in orbit around Earth since 1990. It has accomplished a very long list of discoveries, including observing celestial objects in unprecedented detail. Hubble has observed and recorded evidence of ultraviolet light emission (blocked by Earth's atmosphere), the formation of young stars, early planet formation, supermassive black holes, indirect evidence for exoplanets, and an accelerated expansion of the Universe [8].

A final word about telescopes. Even though many telescopes collect and focus light, others can use digital technologies that detect other types of electromagnetic waves, waves that our eyes cannot see. Since different astronomical phenomena produce different types of radiation, it is important to 'see' celestial objects in x-rays (e.g. Chandra Space Observatory), in gamma rays (e.g. Fermi Space Telescope), in ultraviolet (e.g. Hubble Space Telescope, Far Ultraviolet Spectroscopic Explorer), infrared (e.g. Herschel Space Observatory, James Webb Space Telescope—scheduled to launch in 2018), in microwaves (e.g. Wilkinson Microwave Anisotropy Probe) and

Figure 1.6. At the Arecibo Observatory, one of the authors (in a green shirt) and several science teachers from schools in Puerto Rico explore the 900-ton platform that houses instruments for detecting or emitting radio waves. The main telescope's spherical reflector, 305 m in diameter, is about 140 m below the platform.

in radio waves (e.g. Arecibo Radio Observatory, figure 1.6). These data allow scientists to learn much more about space.

1.3 Personal visits

One of the first things scientists needed to accomplish to explore space directly was to temporarily overcome the Earth's gravitational attraction, sending objects and people into orbit. To do this we learned how to design rockets, how to computerize as many processes as possible and, in particular, to understand the long term physiological effects humans might endure if they want to personally explore the Solar System and beyond. For example, we know that low linear energy transfer radiation, galactic cosmic radiation and proton radiation due to solar particle events pose a threat to humans in space, especially if we leave the protection of the Earth's magnetic field [9].

Experiencing weightlessness for more than a few months causes a number of negative physiological effects, including loss of bone and muscle tissue, and vision problems [10, 11]. The isolation and inability to see family or friends might cause as yet unknown psychological effects. From the astronauts that have inhabited the International Space Station, we have learned that an extensive routine of cardio-vascular exercises is essential to maintain muscle tone and bone density. Engineers have adapted exercise bikes, treadmills and resistive exercise machine, among others, so that they can effectively work in a weightless environment (figure 1.7). Astronauts' fitness parameters are carefully monitored and exercise routines are adjusted accordingly.

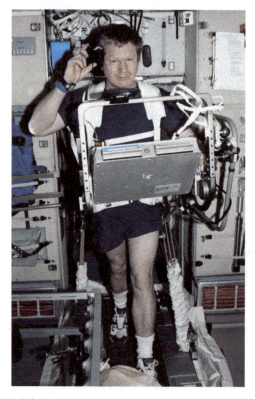

Figure 1.7. Expedition One mission commander William Bill Shepherd is photographed exercising on the ISS treadmill in the Zvezda Service module. Note that the astronaut is strapped to the treadmill; elastic bands to his left and right provide a downward pull equivalent to that of gravity. Photo courtesy of NASA.

Another place that we explored directly was the surface of the Moon. Between July 1969 and December 1972, six Apollo Missions landed on our natural satellite. Although the astronauts completed a number of experiments, scientists were particularly curious to see whether returned samples could shed light on the formation of the Moon. Before the 1960s, several theories were postulated to explain the origin of the Moon [12, 13]:

- Capture: the Moon formed elsewhere in the Solar System, was passing by and was captured by the Earth's gravity.
- Fission: the Earth's rotation flung out the material that would form the Moon.
- Condensation: the Moon and Earth formed at the same time from the same material.
- Collision: two large planetesimals collided, one ejected piece became the Moon and the other became the Earth.
- Ring ejection: a planetesimal the size of Mars collided obliquely with an already-formed Earth; the ejecta created a ring, which coalesced to form the Moon.

An analysis of the physical and chemical composition of the Moon rocks supported the 'ring ejection' theory of moon formation [14].

1.4 Robotic visits

Not only do manned space missions constantly endanger the life of astronauts, they also complicate the design of spacecraft because life-sustaining cargo and systems need to be included. This has resulted in the increasing use of semi-automated probes to explore space, remotely controlled by a crew of scientists, engineers and technicians. It is much easier to design an unmanned space probe.

We have sent unmanned probes to the Sun (Ulysses, SOHO), the Moon (Luna missions, Lunar Reconnaissance Orbiter), the inner planets (Magellan, Venera, Mariner 10, Viking), the outer planets (Galileo, Cassini, figures 1.8 and 1.9; New Horizons), asteroids (NEAR Shoemaker) and comets (Deep Impact, Rosetta). Currently, Voyager 1 and 2, launched in 1977, are reaching beyond the Kuiper Belt into interstellar space. Voyager 1 is more than 130 AU from Earth; Voyager 2 travels in a different direction and it is more than 108 AU from us. One AU, or astronomic unit, is equivalent to 93 million miles or 150 million km.

These space probes include communication systems for sending and receiving data, and a number of instruments that can measure phenomena from electro-magnetic waves to magnetic and gravitational fields. Optical cameras allow scientists to take stunning pictures in unprecedented detail.

Figure 1.8. In 1992, the Galileo spacecraft returned color-enhanced images of the Moon. Photo courtesy of NASA.

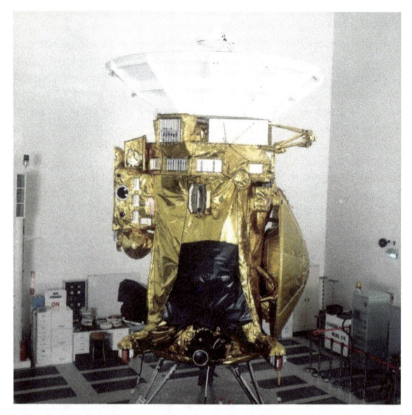

Figure 1.9. The Cassini spacecraft, before it was launched in 1997. It was able to perform 27 different investigations to probe the mysteries of the Saturn system. Photo courtesy of NASA.

Scientists have learned a lot from these missions. For example, recent data from the Voyager probes suggest that the Sun's zone of influence, called the heliosphere, is not symmetric. This revealed a new understanding of the interaction between the heliopause—the boundary between the Sun's influence and the interstellar medium— in the form of termination shocks and plasma flows [15].

The exploration of Mars using rovers such as Spirit and Opportunity have revealed that the planet is currently inhospitable to life. Scientists now know that it once had a climate that could have supported life billions of years ago, that briny liquid water might exist closer to the surface than previously thought, that organic molecules are present in some of Mars' soil and that the cosmic radiation dose from space at ground-level is about 600 times that of Earth's surface [16–18].

In November 2014, the European Space Agency's Rosetta mission's Philae lander touched down on the surface of comet 67P/Churyumov–Gerasimenko (figure 1.10). Here we discovered, among other things, that the majority of outgassing activity from the comet is occurring near the middle of the dumbbell-shaped comet, that its surface is more hydrocarbon-rich, complex and varied that previously thought, and that a cloud of particles orbit the comet despite its minute gravitational attraction [19–22].

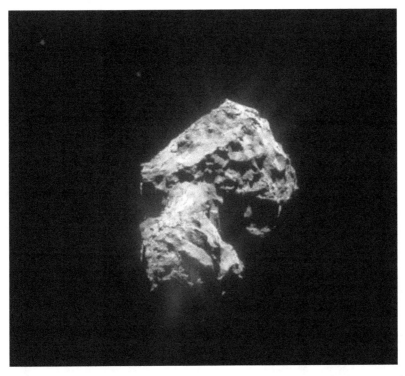

Figure 1.10. Comet 67P/Churyumov–Gerasimenko as seen from the Rosetta navigation camera image on May 20, 2015. Rosetta is about 163.6 km from the comet center. The image has a resolution of 13.9 m/pixel and measures 14.3 km across. Photo courtesy of ESA/Rosetta/NavCam.

In this chapter, the reader became familiar with why humans feel compelled to explore the unknown, what instruments we use to learn from the electromagnetic waves we receive on Earth from outer space, and how humans have sent manned and unmanned explorers to collect even more data. The next chapter focuses on the exploration of a specific type of astronomical object, planets, both inside and outside the Solar System.

References

[1] Ball P 2012 *Curiosity: How Science Became Interested in Everything* (Chicago, IL: University of Chicago Press)
[2] Silvia P J 2006 *Exploring the Psychology of Interest* (New York: Oxford University Press)
[3] Firestein S 2012 *Ignorance: How it Drives Science* (New York: Oxford University Press)
[4] Pisula W 2009 *Curiosity and Information Seeking in Animal and Human Behavior* (Boca Raton, FL: Brown Walker)
[5] Fowler H 1965 *Curiosity and Exploratory Behavior* (New York: Macmillan)
[6] Tyson N D 2006 Delusions of space enthusiasts *Nat. Hist.* **115**(9) 21–25
[7] Zotti G 2008 Tangible heritage: production of astrolabes on a laser engraver *Comput. Graph. Forum* **27**(8) 2169–77
[8] Witze A 2015 Biography of a space telescope: voices of Hubble *Nature* **520**(7547) 282–6

 [9] Chancellor J C, Scott G B I and Sutton J P 2014 Space radiation: the number one risk to astronaut health beyond low Earth orbit *Life* **4** 491–510

[10] Mader T H *et al* 2011 Optic disc edema, globe flattening, choroidal folds, and hyperopic shifts observed in astronauts after long-duration space flight *Ophthalmology* **118**(10) 2058–69

[11] Stevens J 2004 Bumpy road to Mars *Smithsonian* **35**(3) 39–43

[12] Daly R A 1946 Origin of the Moon and its topography *Proc. Am. Phil. Soc.* **90**(2) 104–19

[13] Jackson A P and Wyatt M C 2012 Debris from terrestrial planet formation: the Moon-forming collision *Mon. Not. R. Astron. Soc.* **425**(1) 657–79

[14] Mason B 2002 Mystery of Moon's origins solved *New Sci.* **176**(2374/2375) 15

[15] Richardson J D 2013 Voyager observations of the interaction of the heliosphere with the interstellar medium *J. Adv. Res.* **4**(3) 229–33

[16] Freissinet C *et al* 2015 Organic molecules in the Sheepbed Mudstone, Gale Crater, Mars *J. Geophys. Res.: Planets* **120**(3) 495–514

[17] Hassler D M *et al* 2014 Mars' surface radiation environment measured with the Mars Science Laboratory's Curiosity Rover *Science* **343**(6169) 1244797

[18] Martin-Torres J F *et al* 2015 Transient liquid water and water activity at Gale crater on Mars *Nat. Geosci.* **8**(5) 357–61

[19] Capaccioni F *et al* 2015 The organic-rich surface of comet 67P/Churyumov–Gerasimenko as seen by VIRTIS/Rosetta *Science* **347**(6220) aaa0628

[20] Rotundi A 2015 Dust measurements in the coma of comet 67P/Churyumov–Gerasimenko inbound to the Sun *Science* **347**(6220) aaa3905

[21] Sierks H *et al* 2015 On the nucleus structure and activity of comet 67P/Churyumov–Gerasimenko *Science* **347**(6220) aaa1044

[22] Thomas N *et al* 2015 The morphological diversity of comet 67P/Churyumov–Gerasimenko *Science* **347**(6220) aaa0440

Chapter 2

The ABC of exoplanets

2.1 What is a planet?

Before 2006, there was no formal definition of a 'planet'. Scientists had several guidelines, though. A celestial object could be considered a planet if it was relatively large, mostly spherical in shape, and if it orbited around a star, like the Sun. For astronomers, it was relatively straightforward to determine if a celestial object was round and if it orbited around the Sun. The third guideline, regarding the size of the celestial body, was the trickiest one to pin down.

Consider Ceres, at 938 km in diameter and located in the Mars–Jupiter Asteroid Belt (figure 2.1). In 1801 it fit quite well the three guidelines for what a planet should look like. Ceres and a few of its large companions were labeled as planets for many decades after their discovery [1]. However, as many other objects were discovered in that zone, Ceres' classification was changed to that of a large asteroid. Note that the push for changing Ceres' classification was not due to Ceres itself, but to the discovery of many similar objects in its vicinity. Instead of adding more planets to the list, astronomers decided to classify objects orbiting the Sun between Mars and Jupiter as asteroids.

Something similar happened to Pluto and its classification as a planet (figure 2.2). Discovered by Clyde Tombaugh in 1930, at 2370 km in diameter, Pluto was considered a planet from 1930 to 2006 [2]. That is, until Eris showed up. This celestial body was discovered in 2005 by Caltech astronomer Mike Brown and his team [3] at a location about three times more distant from the Sun than Pluto.

Eris is roughly the same size as Pluto. Further data analysis suggested that Eris has more mass than Pluto. As scientists pondered the possibility of finding even more celestial bodies like Eris in this region of the Solar System (known as the Kuiper Belt), they faced a dilemma. In the same way that Ceres was eventually not considered a planet anymore, could Pluto, Eris and other yet to be discovered celestial bodies no longer be considered planets, becoming celestial bodies among many in the Kuiper Belt?

doi:10.1088/978-1-6817-4401-8ch2   2-1   © Morgan & Claypool Publishers 2016

Figure 2.1. Close-up image of dwarf planet Ceres, taken by NASA's Dawn spacecraft on 6 June 2015. The bright spots are likely salt or ice deposits. Photo courtesy of NASA/JPL-Caltech/UCLA/MPS/DLR/IDA.

Figure 2.2. On 14 July 2015, the telescopic camera on NASA's New Horizons spacecraft took some of the highest resolution images of Pluto. Photo courtesy of NASA/JHUAPL/SwRI.

In 2006, a majority of astronomers attending the International Astronomical Union's General Assembly in Prague agreed to this official definition of a planet: 'A planet is a celestial body that (a) is in orbit around the Sun, (b) has sufficient mass for its self-gravity to overcome rigid body forces so that it assumes a hydrostatic equilibrium (nearly round) shape, and (c) has cleared the neighborhood around its orbit' [4]. This definition currently limits the number of planets to eight: Mercury, Venus, Earth, Mars, Jupiter, Saturn, Uranus and Neptune.

What about Pluto, Eris and others? At the same International Astronomical Union meeting, astronomers created a new definition, the 'dwarf planet'. This new type of celestial object fits criteria (a) and (b) above, but not criterion (c), that is, they are not large enough to clear the neighborhood of their orbits. The official list of dwarf planets now includes Pluto, Eris and Ceres, as well as the recently discovered Haumea and Makemake. About half a dozen more celestial bodies are on a 'waiting list' to be officially declared dwarf planets, including V774104, a trans-Neptunian object discovered in November 2015 which appears to be even farther away than Eris [5].

2.2 What is an exoplanet?

By applying the previous definition of a planet, we can define what an exoplanet is. The Greek prefix 'exo-' means 'outside' or 'external', meaning that an exoplanet is a massive, round celestial body in orbit around a star other than the Sun. Note that the third part of the definition of a planet is, at this point, very difficult to verify. Although our technology is getting good enough to detect exoplanets, we are still far away from detecting whether smaller celestial bodies have been cleared out of an exoplanet's orbit.

Astrophysicists Aleksander Wolszczan and Dale Frail discovered the first three exoplanets back in 1992. Since then, and as of December 2015, about 2000 exoplanets have been discovered and confirmed (figure 2.3), and about 4700 more are considered 'candidates', that is, the data need to be analyzed further to verify their accuracy.

Figure 2.4 shows the radius and orbital period distributions of candidate and confirmed exoplanets as of December 2015. For comparison purposes Earth, represented by the green square, would be located at the intersection of the 365-day period (x-axis) and 1 Earth radii (y-axis).

This graph is interesting because it demonstrates that most of the known exoplanets are larger than Earth and have a shorter orbital period, that is, they are closer to their star than the Earth is to the Sun. It is important to note that this might not be an accurate representation of the actual distribution of real exoplanets, but more of a systematic bias caused by our technical limitations. In other words, we have found the 'low-hanging fruit', very large planets that orbit very fast around their parent star. Using our Solar System as an example, it is plausible that many exoplanets smaller than Earth do exist at much larger distances away from their stars. Regardless, just the fact that we are starting to discover and describe the physical characteristics of planets outside our Solar System, is considered by many as one of the greatest achievements of science [6].

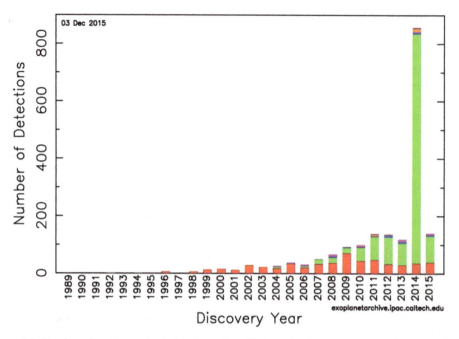

Figure 2.3. Number of yearly exoplanet detections since 1989. Each color represents a different exoplanet detection method. Image courtesy of the NASA Exoplanet Archive, NASA/JPL-Caltech.

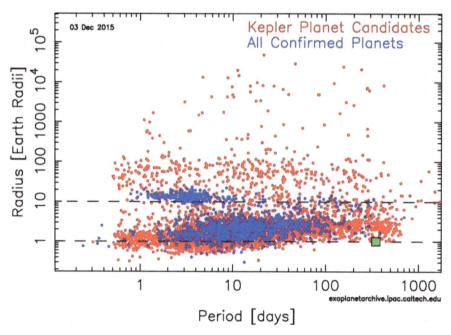

Figure 2.4. Each dot represents either a candidate exoplanet or a confirmed one. Most of the currently known exoplanets are larger and faster than Earth. Image courtesy of the NASA Exoplanet Archive, NASA/JPL-Caltech.

Exoplanets have been classified by size according to a scale based on our own Solar System. Those that appear to be composed mainly of gas are classified as jovian (Jupiter-sized) or neptunian (Neptune-sized). Exoplanets that appear to be composed mostly of rock are classified as superterran (if they have between 5–10 times Earth's mass or between 1.5–2.5 times Earth's diameter), terran (if they have between 0.5–5 times Earth's mass or between 0.8–1.5 times Earth's diameter), or subterran (if they have between 0.1–0.5 times Earth's mass or between 0.4–0.8 times Earth's diameter).

2.3 Detecting exoplanets

Detecting planets orbiting other stars is very difficult [7] and distance is a crucial obstacle. For example, our Sun is about 8.32 light-minutes from Earth, that is, light takes 8.32 minutes to reach Earth once it leaves the surface of the Sun. The light emitted by Proxima Centauri, the closest star to Earth, takes about 4.24 light-years. This is almost 270 000 times the distance! Our current limit on how far we can detect exoplanets is around 13 000 light-years away [8].

A second obstacle to detecting exoplanets is that, unlike stars, exoplanets do not emit visible light on their own. Against the blackness of space, the planets are simply invisible to the naked eye. Fortunately, some of them emit infrared waves, either because the planet is warm or because it reflects waves from its central start, which may be detected. Other reflect faint but visible light from their central star (figure 2.5).

Figure 2.5. A waning crescent Moon is featured in this image photographed by an Expedition 24 crew member on the International Space Station. Note that the section of the Moon that is not directly reflecting sunlight is indistinguishable from the black background. Photo courtesy of NASA.

Figure 2.6. From the International Space Station, Expedition 42 Flight Engineer Terry W Virts took this photograph of the Gulf of Mexico and US Gulf Coast at sunset. The Sun's brightness makes it impossible to see the faint sunlight reflected by planets Mercury and Venus. Photo courtesy of NASA/Terry Virts.

The third hurdle to finding exoplanets is that a star is much brighter than any light reflected by the exoplanet (figure 2.6). Consider, for example, the planet closest to our own Sun, Mercury. Even professional astronomers have a hard time finding Mercury in the night sky; they know that the best time for locating Mercury with a telescope is to wait several minutes after sunset, so that the Sun's brightness is greatly diminished. However, if they wait too long, Mercury will move below the horizon.

Despite these seemingly insurmountable limitations, scientists have developed several direct and indirect techniques to find exoplanets [9, 10]. Let's look at some of the most promising ones.

Doppler shift. Stars produce light because of the nuclear fusion reactions that take place inside their cores. This nuclear energy transforms into electromagnetic waves of different frequencies. The temperature of a star directly correlates with the peak frequency or color that we can see from Earth, that is, as long as the star is not moving with respect to Earth. In other words, a star without any exoplanets will always show the peak light frequency in exactly the same location in a spectrograph.

However, when a star has exoplanets orbiting it, both the star and the exoplanets move around the system's center of mass. If the plane of the star system is more or less parallel to Earth, from our planet the star sometimes moves toward Earth and sometimes moves away from Earth. This relative motion shifts the peak color frequency of the star ever so slightly over time, closer to red when the star moves away and closer to blue when the star moves toward Earth (figure 2.7). Using a Doppler effect equation, changes in wave frequency can be converted to velocity. The relative velocity curve produced based on Doppler shift data is very important since it allows scientists to estimate the mass and distance of the star's exoplanet.

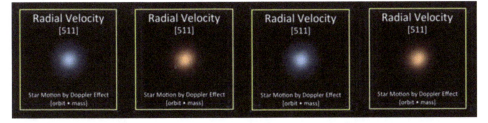

Figure 2.7. This is a visualization of the Doppler shift exoplanet detection technique. It represents the visual appearance of a star as it is affected by the presence of planets. A star's color will vary depending on whether it is moving away (red-shift) or towards our planet (blue-shift). The effect of the planets on their stars are very exaggerated for clarity since they are too weak to be visible by the naked eye. Image courtesy of PHL @ UPR Arecibo.

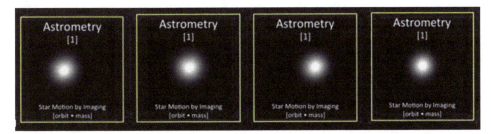

Figure 2.8. This is a visualization of the astrometry exoplanet detection technique. It represents the visual appearance of a star as it is affected by the presence of planets. When the star location is to the left, it means the exoplanet is located to its right, and vice versa. The effect of the planets on their stars are exaggerated for clarity since they are too weak to be visible by the naked eye. Image courtesy of PHL @ UPR Arecibo.

Astrometry. Since the Greek prefix 'astro-' means 'star' and the Greek suffix '-metry' means 'to measure', it is easy to remember that astrometry is the measurement of the positions and motions of the celestial bodies. A star without exoplanets will always be located exactly in the same celestial coordinates in the sky, relative to Earth. But a star with large exoplanets, as mentioned previously, will move around the system's center of mass. If the plane of the star system is more or less perpendicular to Earth, the star will move ever so slightly around a central point, suggesting the influence of exoplanets (figure 2.8).

Transit. Occasionally, a celestial body appears to cross in front of its star. This event is known as a transit. Depending on the relative location of Earth, a celestial body can cover most of the star, like when the Moon appears to cover the Sun and it gets noticeably dark on our planet. On the other hand, you can get a planet transit that is barely noticeable. Figure 2.9 shows the planet Venus as it transits in front of the Sun. From Earth, the Sun's diameter is much larger than the apparent diameter of Venus.

Exoplanets can also appear to cross between their central star and Earth. The change in the star's illumination cannot be perceived except with the most potent telescopes and computers, and only when the exoplanet's orbit is just right, as seen from Earth (figure 2.10). Despite the fact that the alignment of Earth, the star and its

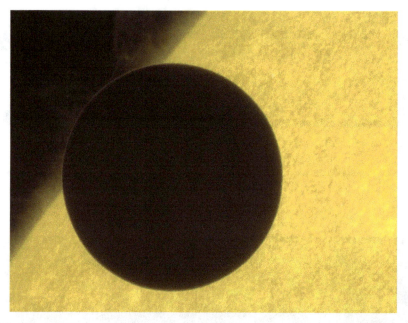

Figure 2.9. Occasionally, the planet Venus crosses in front of the Sun. Using sophisticated instruments, scientists can measure the change in light intensity as a transit occurs, even in stars far away from the Solar System. Photo courtesy of JAXA/NASA/Lockheed Martin.

Figure 2.10. This is a visualization of the transit exoplanet detection technique. It represents the visual appearance of a star as it is affected by the presence of planets. When a planet crosses in front of a star, its brightness decreases ever so slightly, before returning to the original brightness. The effect of the planets on their stars are exaggerated for clarity since they are too weak to be visible by the naked eye. Image courtesy of PHL @ UPR Arecibo.

exoplanet must be almost perfect, this method has produced the largest number of exoplanet discoveries. It is plausible that many more thousands of star systems have an alignment with Earth that is slightly off, making it impossible for scientists to see their exoplanets with this method.

Direct imaging. Current technology allows astronomers to directly see exoplanets [11]. Basically, the light from a star must be physically or digitally dimmed or masked so that the exoplanet's reflected light or emitted infrared radiation can be detected by the instruments (figure 2.11). This method, so far, only applies to those Jupiter-like planets that are relatively far away from their star (and are not occulted

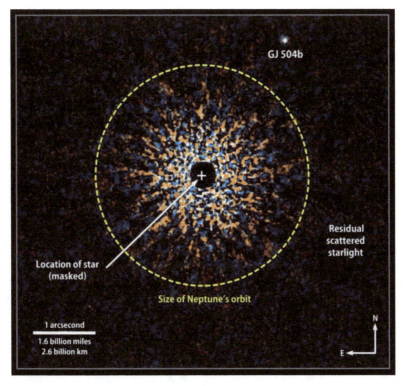

Figure 2.11. This composite combines Subaru images of GJ 504 using two near-infrared wavelengths. Once processed to remove scattered starlight, the images reveal the orbiting planet, GJ 504b. Photo courtesy of NASA's Goddard Space Flight Center/NAOJ.

by the much brighter star) and are quite young and hot. Direct detection has the advantage that some of the exoplanet's physical parameters can be calculated with greater precision.

Two other methods for detecting exoplanets include pulsar timing and gravitational microlensing. In the case of *pulsar timing*, as stars rotate, they emit electromagnetic radio pulses at a clock-like regular interval. If those stars have exoplanets, the pulses briefly go out of synchronization. Comparing the disturbed and undisturbed timing of the radio pulses allows scientists to infer the exoplanet's mass and orbital period. The second method, *microlensing*, takes advantage of Einstein's general theory of relativity. In this case, light from a star behind an unknown exoplanet is slightly deviated by its mass. To be clear, this not a case of a planet blocking the light but bending it, creating the illusion that the background star briefly moved from its normal location.

We have learned quite a lot! The occasionally fluctuating concept of a planet was defined and applied to planets way beyond our Solar System. Also, a variety of direct and indirect methods for detecting exoplanets were described. Our next stop on the road to discover habitable worlds is to review the concept of life and introduce the concept of habitability.

References

[1] Loomis E 1854 The zone of small planets between Mars and Jupiter *Ninth Annual Report of the Board of Regents of the Smithsonian Institution* (Washington, DC: AOP Nicholson) 137–46

[2] Soter S 2007 What is a planet? *Sci. Am.* **296**(1) 34–41

[3] Brown M E, Trujillo C A and Rabinowitz D L 2005 Discovery of a planetary-sized object in the scattered Kuiper Belt *Astrophys. J. Lett.* **635**(1) L97–L100

[4] Ekers R 2006 *IAU Planet Definition Committee, Dissertatio CVM Nuncio Sidereo III* 16-8 4–5

[5] Beatty K 2015 V774104: Solar System's most distant object *Sky Telesc.* November 21, http://www.skyandtelescope.com/astronomy-news/v774104-most-distant-solar-system-object-11212015/

[6] Sergison D 2013 High precision photometry: detection of exoplanet transits using a small telescope *J. Br. Astron. Assoc.* **123**(3) 153–6

[7] Jonathan L I, Macintosh B and Peale S 2009 The detection and characterization of exoplanets *Phys. Today* **62**(5) 46–51

[8] Yee J 2015 NASA's Spitzer spots planet deep within our Galaxy http://www.spitzer.caltech.edu/news/1746-feature15-05-NASA-s-Spitzer-Spots-Planet-Deep-Within-Our-Galaxy

[9] Akeson R L *et al* 2013 The NASA exoplanet archive: data and tools for exoplanet research *Publ. Astron. Soc. Pac.* **125**(930) 989–9

[10] Angerhausen D, Krabbe A and Iserlohe C 2010 Observing exoplanets with SOFIA *Publ. Astron. Soc. Pac.* **122**(895) 1020–9

[11] Rice K 2014 The detection and characterization of extrasolar planets *Challenges* **5** 296–323

Chapter 3

When is a planet habitable?

3.1 What is life?

It is very easy to convince a person that the chicks shown in figure 3.1 are alive. They move, chirp, eat, react to their environment (temperature, sounds, brightness, etc), among other cute things chicks do. But, is the cotton towel under the chicks alive? Cotton grows on plants, and we know plants are alive, right? They do not move around by themselves or chirp like the chicks, though. Is pollen alive? Are fungi alive? Are bacteria alive? Are viruses alive? This issue of classifying living and non-living things can get quite complicated.

Before learning about what makes a planet habitable, it is important to think about what might inhabit them. Scientists have compiled a few simple definitions of what life is [1]. Some of these definitions include:

- 'Life is self-reproduction with variations' [2].
- 'Life is a self-replicating, evolving system based on organic chemistry' [3].
- 'Living things are systems that tend to respond to changes in their environment, and inside themselves, in such a way as to promote their own continuation' [4].
- 'Life is a self-sustained chemical system capable of undergoing Darwinian evolution' [5].
- 'Living organisms are autopoietic systems' [6].

Some common elements of life can be inferred from these statements. Living things sustain themselves chemically, self-reproduce and are capable of evolution. Additional characteristics of organisms include their ability to feed from, and respond to, the environment. Koshland [7] synthesized these and other ideas about life in his 'seven pillars of life': a program (an organized plan); improvisation (a way to change the program); compartmentalization (a means of separating self from the outside world); energy (to fuel chemical reactions); regeneration (to repair and replace itself); adaptability (to respond to the environment); and seclusion (to insulate chemical reactions from one another).

doi:10.1088/978-1-6817-4401-8ch3   3-1   © Morgan & Claypool Publishers 2016

Figure 3.1. Chicks are obvious examples of living organisms. Could other planets be able to support a chicken population?

Figure 3.2. Is this a microscopic jellyfish, or a diatom, or some shelled, bioluminescent organism of the deep? No, it is the Cat's Eye Nebula, located 3000 light-years from Earth. Will humans be able to distinguish between living and non-living organism outside Earth? Photo courtesy of NASA, MAST, STScI, AURA and Vicent Peris (OAUV/PTeam).

Of course, the previous definitions of life are based on our observation of what life looks like on Earth (see figure 3.2). Scientists have estimated that there are between 7 and 10 million species of multi-cellular organisms on our planet and many more prokaryotic, or single-celled organisms, most of which are still unknown to science [8].

Phylogenetic Tree of Life

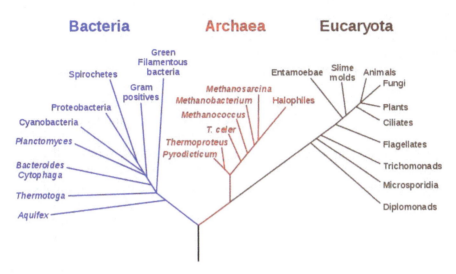

Figure 3.3. Phylogenetic tree, or the 'tree of life'. Note that all organisms emerged from a single type of ancestor. Image courtesy of NASA/Eric Gaba.

Despite this breathtaking diversity, all organisms on Earth have evolved over billions of years from a few ancestors (figure 3.3). For example, humans share approximately 90% of their genes with chimpanzees, 84% with dogs, 69% with platypi, 47% with fruit flies, 38% with round worms, 24% with wine grapes, 18% with baker's yeast and 7% with bacteria [9].

Even though it looks as if life on Earth is extremely diverse, different organisms have many things in common. To what extent this knowledge can be extrapolated to planets other than Earth is a matter of debate in the scientific community. Some scientists argue that there are large gaps in our knowledge of Earth's complex history of habitability [10] and that planets need a very specific mix of chemical compounds and physical conditions for life to begin. Others point out that, given the 200–400 billion stars with planetary systems in our Milky Way Galaxy and countless galaxies like it, it is statistically likely that there is a large number of planets with the specific conditions needed to support life.

To explore whether extrasolar planets might be suitable for life or not, scientists need to move beyond the debate just discussed and settle on a paradigm. A paradigm is a scaffolding of ideas, theories and concepts that enables scientists to make sense of newly discovered information [11]. Paradigms are necessary because facts require a context for interpretation. Most scientists use life examples from Earth and the consistency of the laws of chemistry and physics across the Universe as a guiding paradigm, that is, a standard for comparison to make sense of other planets' chances for life. It might not be a perfect paradigm (in fact, it could be far off), but it is the best scientific strategy given what we know about the nature of the Universe [12].

3.2 The concept of habitability

Taking Earth as an example, it is obvious that not everywhere on Earth is equally habitable. Some regions are located at extremes elevations, where the air pressure is one-third that of sea level atmosphere, while other regions are located below the world's oceans, at depths of 36 000 ft (10 000 m) and crushing pressures of 1000 times standard atmospheric pressure. Temperature-wise, Death Valley, in California, can reach temperatures of 130 °F (54 °C), while regions in Antarctica can reach −129 °F (−85 °C). From deserts (see figure 3.4) and polar regions, to rain forests and marine environments, there is an obvious habitability gradient, from worst to best for life.

A habitable environment is one that might support some forms of life, but not necessarily one with life. Habitability or 'habitat suitability' is defined as the suitability of an environment for life. This definition has three components: an environment, a life and a suitability component.

The *environment component* is a description of the physical, chemical, or even biological location of life under consideration, the habitat. It is constrained by space and time limits. This is the astronomy, planetary science, or geology part of the definition. The *life component* requires the selection and knowledge of an individual species or community as the test subject for the habitat. Therefore, given some habitat any habitability measure is always relative to the species or community under consideration. The *suitability component* is the tricky part because it defines the connection between the environment and life.

Potential habitable exoplanets are those extrasolar planets that might be able to support any form of life, from simple life (micro-organisms) to complex life

Figure 3.4. The Great Sandy Desert, in northwest Australia. Each sand dune is 25 m high and 0.5–1.5 km apart. Deserts like this are some of the least habitable places on Earth, yet microscopic and macroscopic life thrive, adapted to the harsh environment. Photo courtesy of NASA.

(plants and animals). For example, billions of years ago, Earth was only able to sustain microbial life; today it can also support plants, animals and a more diverse biosphere. If a planet is potentially habitable, it does not mean that it is necessarily inhabited, only that it has some requirements that scientists consider as necessary for life as we know it, including:

Water. One of these requirements is the presence of water, preferably liquid, on an exoplanet's surface. Water is just a chemical compound that might be necessary for life, but does not necessarily indicate the presence of life. On Earth, water can be found in all three states of matter. About 97.4 % of Earth's surface water is in the form of saltwater, mainly in the ocean. The remaining 2.5% of freshwater is mostly locked in glaciers and ice caps (68.7%; figure 3.5), and groundwater (30.1%). The rest is surface water, either in solid or liquid form [13].

Too little or too much water is not good for life. Terrestrial deserts are characterized by limited water and usually appear devoid of life in the most extremes cases. Microbial life is present even under such conditions, but the absence of plants and animals make them look lifeless. The same is true for many parts of the oceans; there is plenty of water but some deep regions look like terrestrial deserts, without visible life. Most of the complex life in oceans prefer the shallow coastal zones. Even in these cases, microbial life is always present.

Temperature. A second requirement for life, based on the current paradigm, is a relatively moderate temperature. An exoplanet's surface temperature is related to factors such as its distance to the parent star and the star's luminosity. So far, this is the easiest method to assess the potential for life on exoplanets because it can be estimated from direct or indirect observations (figure 3.6).

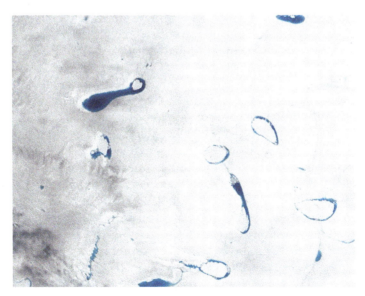

Figure 3.5. Each spring and summer, sapphire-colored ponds spring up, like swimming pools, on the Greenland ice sheet. Some of these are large enough to be visible from space. Photo courtesy of NASA.

Thermal Planetary Habitability Classification

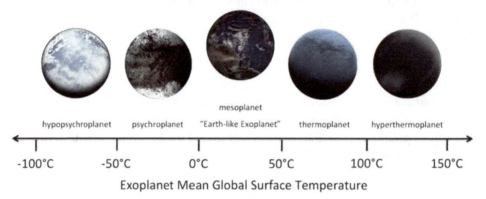

Credit: 2011, PHL @ UPR Arecibo (phl.upra.edu)

Figure 3.6. A proposed thermal planetary habitability classification for exoplanets. We still do not know the actual surface temperature of any exoplanets but classifications like this might be used in the future. Image courtesy of PHL @ UPR Arecibo.

What is a 'relatively moderate' temperature? On Earth, microbial life such as bacteria and archaea has a wide thermal tolerance and growth has been measured at temperatures from −15 °C to about 120 °C. Macrobial life such as animals (metazoa) shows a more restrictive tolerance usually between 0 °C and 50 °C. Most plants are particularly efficient at temperatures close to 25 °C. Water is liquid between 0 °C and 100 °C at standard atmospheric pressures.

Borrowing from a well-known thermal classification for microbial life, where organisms are classified as mesophiles (if they prefer moderate temperatures), psychrophiles (if they prefer cold temperatures) and thermophiles (if they prefer hot temperatures); exoplanets can be similarly classified. In this case, we can talk about mesoplanets, psychroplanets and thermoplanets. All three classes represent potential habitable exoplanets based on their mean global surface temperature. Only mesoplanets (temperate) correspond to Earth-like planets, which may be potentially habitable by complex life as we know it (such as plants and animals). Psychroplanets (cold) and thermoplanets (hot) may only be habitable for microbial life.

Magnetic field. Exoplanets could end up dry and unsuitable for life depending on how they evolve with their star. Therefore, it is necessary to understand the long-term interactions between planets and their star to recognize and characterize habitable worlds. If the conditions in the exoplanet's core are adequate, and if it has a fast enough rotation on its axis, the exoplanet can develop a magnetic field or magnetosphere to surround it (figure 3.7).

Stars produce the energy to maintain the temperate environment of a potentially habitable exoplanet. They also emit harmful energy that could strip atmospheres or damage life at a cellular level. Earth's magnetic field prevents most of the harmful radiation and particles emitted by the Sun from reaching the ground. Without this invisible protection, life on Earth, or at least on its surface, might eventually disappear.

Figure 3.7. Artistic representation of the magnetic field, illustrated with red lines, around a potentially habitable world. Photo courtesy of PHL @ UPR Arecibo.

On Mars, we have a perfect example of what can happen to a planet with a missing magnetic field. Many scientists think that Mars once had significant atmospheric gases and a protective magnetic field. As this planet's magnetosphere dissipated, the solar wind blew the atmosphere off, leaving Mars as we see it today [14, 15]. Similarly, without a magnetic field, it would be unlikely for an exoplanet to keep an atmosphere suitable for life for very long, in particular given the fact that many stars are much more active than our Sun.

Star type. In the 1920s, scientists developed a way to classify stars based on their color, which directly correlates with their surface temperature. For example, Type O stars look blue, corresponding to temperatures of 20 000–35 000 K. On the other end of the spectrum, Type M stars look red, corresponding to temperatures of around 3000 K. Our Sun, with a surface temperature of about 6000 K, is a Type G star (see figure 3.8).

The surface temperature of a star is also related to the rate at which fusion reactions occur inside it. Type O stars have a high rate of fusion, and complete their life cycle quickly, which might not give enough time for life to evolve in a suitable exoplanet. Scientists think that exoplanets at the right distance from Type G, K and M stars have the best chance of evolving life.

Atmosphere and planet composition. Most life forms require some basic atmospheric ingredients like carbon dioxide and oxygen. These gases thermally escape the gravity of a planet unless it has the right mass, size and temperature. In general, small and hot planets will be airless and those that are large and cold will have dense atmospheres (assuming they are not cold enough for atmospheric collapse). Nitrogen, the next volatile gas after hydrogen and helium, is usually neglected in these considerations. Nitrogen is necessary for all life and is mostly made available

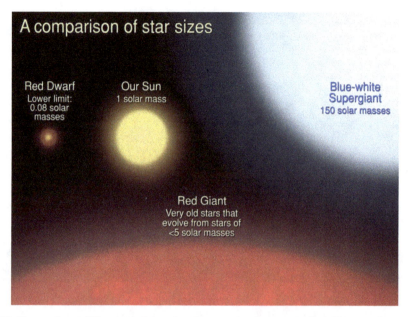

Figure 3.8. Stars evolve at different speeds based on how massive they are. A blue-white supergiant might not last more than 100 million years. Our Sun has a life expectancy of about five more billion years (it is already about five billion years old, half way through its life cycle). A red dwarf might last hundreds of billions of years. Image courtesy of the Astrophysics Science Division (ASD), NASA/GSFC.

to plants by micro-organisms in a process known as nitrogen fixation. Without atmospheric nitrogen (and some is available in rocks, too) the food chain would break down and a planet would probably be only habitable for microbial life.

Scientists can use the thermal escape limits of several gases to define the planetary mass–radius boundaries for atmospheres of habitable planets. Planets able to hold atomic nitrogen will also be able to hold all the other heavier volatiles necessary for life such as carbon dioxide, oxygen, methane, ammonia and water. If a planet has the ability to hold atomic hydrogen, the most abundant element in the universe, then it would probably have a dense, high-pressure atmosphere. This type of planet is probably non-habitable.

Similarly, it is important to measure how compatible for life is the bulk composition of an exoplanet within the habitable zone. Life requires a variety of elements from volatile compounds to iron. Habitable planets require a mixture of these that is readily available. For example, iron planets will have less water and rock, and water or gas planets less iron and rock. Gas planets are probably unsuitable for life; therefore only exoplanets with a rocky composition are potentially habitable.

By combining the atmospheric and bulk composition of potentially habitable planets, a graph like the one in figure 3.9 can be produced. Only exoplanets with the right bulk composition and atmosphere might be habitable. And even then life is not guaranteed. For example, Venus lands in the optimal atmosphere/composition zone, but its temperature is far too high to sustain life as we know it.

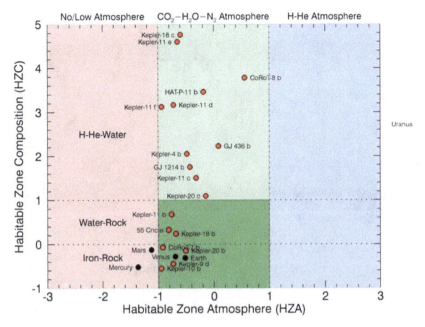

Figure 3.9. This plot combines analyses based on habitable zone atmospheres (*x*-axis) and habitable zone composition (*y*-axis). The optimal parameters for life are shown in dark green. Exoplanets are shown as red dots; Solar System planets are shown as black dots. Courtesy of PHL @ UPR Arecibo.

3.3 Measuring habitability

Habitability metrics are emerging fields within the broader field of astrobiology. As one might expect, habitability metrics originated within the field of ecology and population dynamics to understand the distribution of wild animals and plants (figure 3.10).

Recognizing habitable worlds around other stars is a challenging process. Exoplanets are very far away and most of them are only known by their effect on their parent star, such as periodically making their star wobble or blocking its light. Only a few have been directly imaged, specifically those large and hot enough. Even with these observational limitations scientists are starting to identify potential habitable candidates within the nearly 2000 exoplanets that have been detected so far. The process of identifying habitable worlds can be divided into three steps, as follows.

Physical indicators: basic stellar and planetary properties. Habitable worlds are first recognized by their orbital position with respect to their parent star and some basic planetary properties such as mass and radius. Only exoplanets of the right size and inside the stellar habitable zone, or the right distance from their star to support liquid water, are considered potentially habitable. Small exoplanets will not be able to hold an atmosphere and much larger ones will have very high surface pressures that will even make any water solid, independent of temperature.

Chemical indicators: atmospheric chemistry. Habitable worlds are also recognized by the composition of their atmosphere. A planet that has the right temperatures for liquid water does not necessarily mean that it has water. The light from the atmosphere of the planet is analyzed to search for the presence of water and other

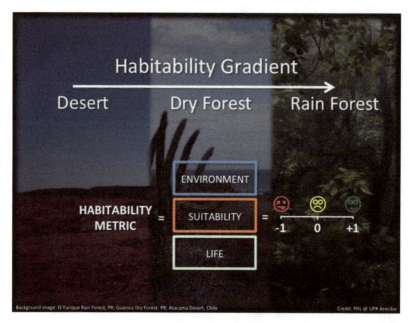

Figure 3.10. Three very distinct habitable environments. Atacama Desert (Chile), Guánica Dry Forest (Puerto Rico) and El Yunque Rain Forest (Puerto Rico). A habitability metric is a number that can compare the physical and chemical properties of an exoplanet with that of Earth. Courtesy of PHL @ UPR Arecibo.

gases required by life such as oxygen, carbon dioxide, methane and nitrogen. The presence of these gases is not a guarantee for the existence of life but indicates potential habitability.

Biological indicators: biosignatures. Habitable or inhabited worlds are finally confirmed by the observation of strong indicators for the presence of life. Life that is abundant and globally distributed will have an impact on the chemistry of the atmosphere and how light is absorbed at the planetary surface. Any vegetation will absorb light at some particular energies that can be recognized. Oxygen is a very reactive gas and tends to be trapped by rocks in a diverse mixture of compounds. The increase of oxygen in Earth's atmosphere, for instance, was due to microbial life. In fact, today half of the oxygen produced by life is by oceanic phytoplankton, a form of microbial life (the other half is by plants).

These steps are necessary to confirm that any extrasolar worlds are in fact habitable by any terrestrial life or perhaps even inhabited. Unfortunately, scientists are currently only able to barely complete the first step and in some very special cases the second. Future ground and orbital observatories are necessary to thus confirm the habitability or habitation of exoplanets.

We expect that extraterrestrial life on exoplanets, if any, is most probably microbial life. Even at close range they are not evident, but we know that abundant microbial life can impact the atmosphere of a planet, and that is something we can measure. More complex life such as animals and vegetation is a second possibility. In the case of Earth, in 99.999% of its history it was inhabited by microbial life, animals and plants, and without intelligent life. Intelligent life should be much rarer. Only in the last few hundred thousand years have we had a highly intelligent species on Earth, that is, homo sapiens.

One way to determine whether exoplanets are potentially habitable, is to combine their mean radius, bulk density, escape velocity, surface temperature and similar parameters discussed in this chapter (as well as others beyond the scope of this book) into a single number or a 'score'. The better the score, the better the probability of life on a given exoplanet. One such score, based on our paradigm, is the Earth Similarity Index (ESI).

The ESI is a multi-parameter first assessment of Earth-likeness for solar and extrasolar planets as a number between zero (no similarity) and one (identical to Earth). Such similarity indices are used in many fields and provide a powerful tool for categorizing and extracting patterns from large and complex datasets. The ESI can be used to prioritize exoplanet observations, perform statistical assessments and develop planetary classifications.

For instance, Earth-like planets can be defined as any planetary body with a similar terrestrial composition and a temperate atmosphere. As a general rule, any planetary body with an ESI value over 0.8 can be considered an Earth-like planet. This means that the planet might have a rocky composition and a temperate atmosphere. Planets with ESI values in the 0.6–0.8 range (e.g. Mars) might still be habitable since habitability depends on many other factors.

A final note on habitability. So far this discussion has focused on exoplanets. However, other celestial bodies that are not exoplanets could be potentially habitable. Habitable moons might support a subsurface biosphere by tidal/internal heating [16]. In fact, several lines of evidence suggest the presence of subsurface salty oceans on the icy moons Europa (figure 3.11) and Ganymede (two of Jupiter's moons) and Enceladus (one of Saturn's moons).

Figure 3.11. The fascinating surface of Jupiter's icy moon Europa looms large in this newly reprocessed color view, made from images taken by NASA's Galileo spacecraft in the late 1990s. Photo courtesy of NASA/JPL-Caltech/SETI Institute.

Habitable nomads, free-floating planets ejected from a planetary system, might have subsurface oceans that could be potentially habitable. These nomads are only limited to their internal energy. Unfortunately, it will be very difficult or impossible to assess the subsurface habitability of any exoplanet and we might only be able to measure surface habitability for planets, or those moons in the habitable zone.

References

[1] Prud'homme-Généreux A 2013 What is life? An activity to convey the complexities of this simple question *Am. Biol. Teach.* **75**(1) 53–7

[2] Trifonov E N 2011 Vocabulary of definitions of life suggests a definition *J. Biomol. Struct. Dyn.* **29** 259–66

[3] Pace N R 2001 The universal nature of biochemistry *Proc. Natl Acad. Sci. USA* **98** 805–8

[4] Morales J 1998 The definition of life *Psychozoan* **1** http://baharna.com/philos/life.htm

[5] Joyce G F 1995 The RNA world: life before DNA and protein *Extraterrestrials: Where Are They?* ed B Zuckerman and M H Hart 2nd edn (Cambridge, UK: Cambridge University Press) 139–51

[6] Margulis L and Sagan D 2000 *What is Life?* (Berkeley, CA: University of California Press)

[7] Koshland D E Jr 2002 The seven pillars of life *Science* **295**(5563) 2215–6

[8] Mora C, Tittensor D P, Adl S, Simpson A G B and Worm B 2011 How many species are there on Earth and in the ocean? *PLoS Biol.* **9**(8) e1001127

[9] Herrero J 2013 Genes are us. And them *Natl Geogr.* http://ngm.nationalgeographic.com/2013/07/125-explore/shared-genes

[10] Lunine J I 1999 *Earth: Evolution of a Habitable World* (Cambridge, UK: Cambridge University Press)

[11] Kuhn T 1996 *The Structure of Scientific Revolutions* (Chicago, IL: University of Chicago Press)

[12] Lammer H *et al* 2009 What makes a planet habitable? *Astron. Astrophys. Rev.* **17**(2) 181–249

[13] Gleick P H 1993 *Water in Crisis: a Guide to the World's Fresh Water Resources* (New York: Oxford University Press)

[14] Connerney J E P, Acuña M H, Wasilewski P J, Ness N F, Reme H, Mazelle C, Vignes D, Lin R P, Mitchell D L and Cloutier P A 1999 Magnetic lineations in the ancient crust of Mars *Science* **284**(5415) 794–8

[15] Dehant V *et al* 2012 From meteorites to evolution and habitability of planets *Planet. Space Sci.* **72**(1) 3–17

[16] Matsuyama I 2014 Tidal dissipation in the oceans of icy satellites *Icarus* **242** 11–18

Chapter 4

Cataloguing habitable exoplanets

4.1 The first habitable worlds?

The first big announcement of the discovery of a potentially habitable planet came on 30 September 2010. Gliese 581g was announced by a team led by Steven Vogt from the University of Santa Cruz [1]. They used the radial velocity method and observations from the High Accuracy Radial velocity Planet Searcher (HARPS) instruments to detect this planet. It was an exciting discovery and one of the authors of this book (Méndez) remembers telling everybody in his university department. He even had a hard time sleeping that night, thinking about all the implications of the discovery. Was it really Earth-like?

There were claims dating back to 2007 that other planets from the same system, Gliese 581c and Gliese 581d, were also potentially habitable, but planet 'c' was probably too hot and 'd' too cold to support liquid water [2, 3]. Of this system of up to six planets, Gliese 581g apparently was the best candidate. It had a mass and insolation (light from the star) more similar to Earth (figure 4.1). However, there were serious doubts about the existence of Gliese 581g after its discovery. Other astronomers failed to see the signal of the planet using the same data.

The story of the planets around Gliese 581 and their potentially habitability is a long one. In 2014 it was finally shown that Gliese 581g and Gliese 581d were not real planets but observational artifacts caused by the activity of the star [3]. This was the first planetary system believed to have at least two habitable worlds. Today we think that there are just three planets around Gliese 581, all too hot for life. The detection of smaller planets similar to Earth has proved to be a difficult task.

In 2011, several other potentially habitable planets were announced: HD 85512d in July, Gliese 667Cc in November and Kepler-11b in December. HD 85512d was later found to be too hot [4]. Kepler-11b is probably too large and just a smaller version of Neptune. Only Gliese 667Cc seems like a good candidate even today. The story behind the discovery and announcement of Gliese 667Cc is interesting since it was independently discovered by two competing teams of astronomers. The race to detect Earth-like worlds had just started and it was time to catalog them.

doi:10.1088/978-1-6817-4401-8ch4   4-1   © Morgan & Claypool Publishers 2016

Figure 4.1. The six planet system around the star Gliese 581 as believed to exist in late 2010. Today we known that only the planets 'b', 'c' and 'e' are real planets. Image courtesy of PHL @ UPR Arecibo.

4.2 The habitable exoplanets catalog

There are many catalogs dedicated to tracking the discoveries of exoplanets. The four main ones are the Extrasolar Planet Encyclopaedia (exoplanet.eu), the NASA Exoplanet Archive (exoplanetarchive.ipac.caltech.edu), the Exoplanet Orbit Database (exoplanets.org) and the Open Exoplanet Catalogue (openexoplanetcata-logue.com). Not all list exactly the same number of exoplanets due to minor differences in their object selection criteria. Other catalogs focus on a smaller sample of exoplanets based on some particular point of interest. For example, the Habitable Zone Gallery (hzgallery.org) list only planets in the habitable zone, which includes both small and large planets which are not necessarily habitable. Another example is the Exoplanet Transit Database (var2.astro.cz/ETD) which lists only transiting exoplanets.

The Habitable Exoplanets Catalog (HEC; phl.upr.edu/hec) is the only catalog that lists all the exoplanets that are potentially habitable (figures 4.2 and 4.3). The main criteria for inclusion in the catalog is for planets to be in the habitable zone and having between 0.1–10 Earth masses or 0.5–2.5 Earth radii. This is an optimistic criterion so as to include any possible object of astrobiological interest. There is no guarantee that any of these exoplanets are really habitable. We do not know their bulk composition, atmosphere, or actual surface temperatures. All we know is that

Figure 4.2. These are artistic representations of all the planets around other stars (exoplanets) with any potential to support surface life as we know it. Most of them are larger than Earth and we are not certain about their composition and habitability. They are ranked here from closest to farthest from Earth. This selection of objects of interest is subject to change as new interpretations or astronomical observations are made. Earth, Mars, Jupiter and Neptune are shown for scale on the right. Image courtesy of PHL @ UPR Arecibo.

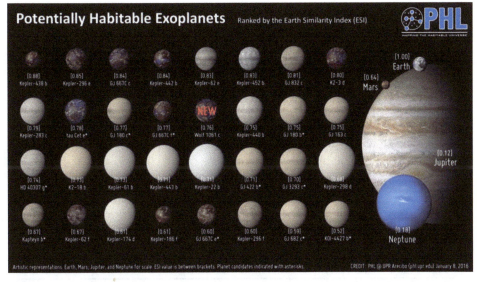

Figure 4.3. These are artistic representations of all the potentially habitable exoplanets ranked from most likely to least likely by the ESI, a measure of Earth-likeness based on stellar flux and planet size. None yet seems to be a true Earth-like planet by this standard (ESI > 0.90). Planets with high ESI values are not necessarily more habitable as habitability depends on other unknown factors such as surface and atmospheric composition. Earth, Mars, Jupiter and Neptune are shown for scale on the right. Image courtesy of PHL @ UPR Arecibo.

they have the potential for surface liquid water, provided that other conditions are right. Planets outside these criteria are not expected to support surface liquid water under any conditions (e.g. gas planets like Neptune or Jupiter).

4.3 The periodic table of exoplanets

The Periodic Table of Exoplanets (PTE) is a way to organize and visualize most of the known exoplanet discoveries, and is divided into six mass/size and three temperatures groups (18 categories; see figures 4.4–4.6). Exoplanets in the hot zone are too close to their parent star to have liquid water. Those in the warm 'habitable' zone have the right distance for liquid water given the right size and atmosphere. Water can only exist as ice for those in the cold zone.

In the PTE divisions miniterrans are low mass bodies, most likely spherical and without atmospheres, similar to Mercury and the Moon. Subterrans are comparable to Mars, terrans to Earth and Venus, and superterrans are a transition group between terrans and neptunians. Neptunians are similar in mass to Neptune and Uranus, and Jovians to Jupiter and Saturn, or larger. Those potentially habitable are described in more detail in the HEC.

The next step in our journey will take us to the details of the potentially habitable worlds that have been detected so far. We will explore their orbits and how they relate to the habitable zone. In our Solar System the planet Venus orbits just outside the inner-edge of the habitable zone while Mars is inside but closer to the outer-edge (figure 4.7). Earth is not at the center of the habitable zone but closer to the

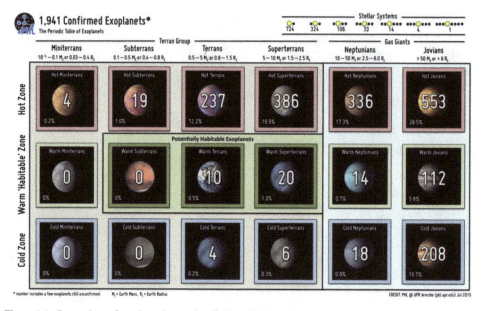

Figure 4.4. Currently confirmed exoplanets classified into 18 thermal-mass categories. The number of exoplanets in each category is shown in the center of each frame and as a percentage in the lower left corner. The diagram also shows the number of multiple stellar systems (top right). The most abundant objects of the confirmed exoplanets are hot Jovians. R_E = Earth radii and M_E = Earth masses. Image courtesy of PHL @ UPR Arecibo.

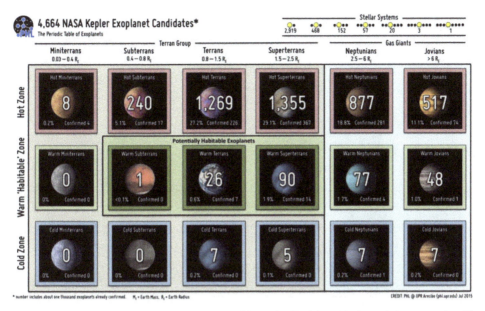

Figure 4.5. Current NASA Kepler exoplanets candidates classified into 18 thermal-size categories. The number of exoplanets in each category is shown in the center of each frame and as a percentage in the lower left corner. The number of those already confirmed are shown in the lower right. The diagram also shows the number of multiple stellar systems (top right). The most abundant objects among the Kepler exoplanets are hot terran and superterrans. R_E = Earth radii. Image courtesy of PHL @ UPR Arecibo.

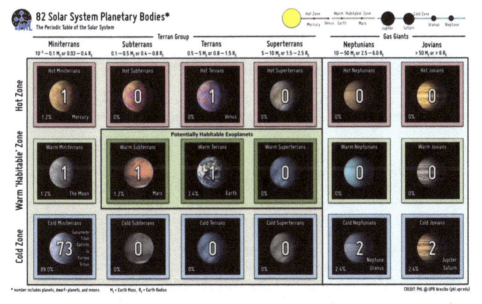

Figure 4.6. Current Solar System spherical objects (with masses over 10^{-5} Earth masses) including planets, dwarf-planets and moons classified into eighteen thermal-mass categories. The number of objects in each category is shown in the center of each frame and as a percentage in the lower left corner. This diagram helps as a basis for comparison with the tables for exoplanets (figures 4.4 and 4.5). The most abundant objects of the Solar System are cold miniterrans (dwarf-planets and moons). R_E = Earth radii and M_E = Earth masses. Image courtesy of PHL @ UPR Arecibo.

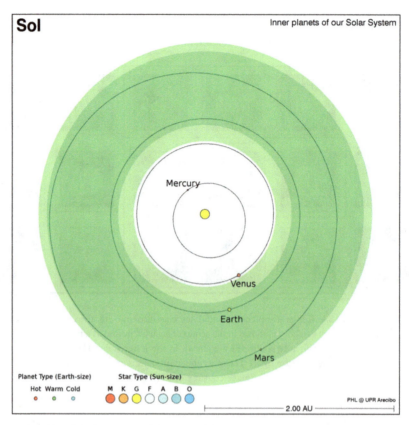

Figure 4.7. The orbits of the inner terrestrial planets of the Solar System. The legend shows for scale purposes the sizes of the planets and stars relative to Earth and the Sun (not to scale with each other or with the orbits). They are color coded to the temperature of the planet and the star. AU = astronomical unit or the average distance between the Earth and the Sun. Image courtesy of PHL @ UPR Arecibo.

inner-edge. It is impossible to tell which orbits are better for habitability since habitability also depends on the planet's composition and atmosphere.

References

[1] Vogt S, Butler R P, Rivera E J, Haghighipour N, Henry G W and Williamson M H 2010 The Lick-Carnegie Exoplanet Survey: a 3.1 $M_\oplus$ planet in the habitable zone of the nearby M3V star Gliese 581 *Astrophys. J.* **723**(1) 954

[2] Udry S *et al* 2007 The HARPS search for southern extra-solar planets *Astron. Astrophys.* **469**(3) L43–7

[3] Robertson P, Mahadevan S, Endl M and Roy A 2014 Stellar activity masquerading as planets in the habitable zone of the M dwarf Gliese 581 *Science* **345**(6195) 440–4

[4] Pepe F *et al* 2011 The HARPS search for Earth-like planets in the habitable zone: I. Very low-mass planets around HD 20794, HD 85512, and HD 192310★★★ *Astron. Astrophys.* **534** A58

Chapter 5

Potentially habitable worlds

5.1 The planets we know by mass

The radial velocity (RV) technique is a very successful method used to detect and measure the mass of exoplanets (chapter 2). It can detect most planets around a star given that they are massive enough to perturb the star and their orbital plane inclination is not perpendicular to the line of sight between the star and us. Unfortunately, the RV technique can only measure the minimum mass of planets, since the inclination is seldom known, and there is no information about the planet's size. This means that the actual mass of planets detected by RV could be much larger.

All potentially habitable worlds known by their mass were discovered using the RV technique with the High Accuracy Radial Velocity Planet Searcher (HARPS) instrument installed at the La Silla Observatory, Chile. HARPS was constructed in 2002 and it is part of the network of telescopes of the European Southern Observatory (ESO). Since this instrument is in the Southern Hemisphere all of the stars it can observe are in the Southern Hemisphere night sky. In 2012 a northern version of HARPS (HARPS-N) was installed at the Roque de los Muchachos Observatory in La Palma, Canary Islands, Spain.

The following paragraphs describe all known stars with planets that might be considered potentially habitable because they have a minimum mass less than 10 Earth masses and orbit within the habitable zone, thus liquid water, if available, might be stable at their surfaces. Planets more massive than 10 Earth masses are expected to be non-habitable gas worlds but there is no guarantee that those below this value are indeed habitable rocky worlds. There is no way to know if they are rocky, water, or gas worlds since we do not know their size to estimate their bulk density. We also do not know their actual surface temperature so some might be hotter or colder than expected depending on the composition of their atmospheres due to potential greenhouse effects.

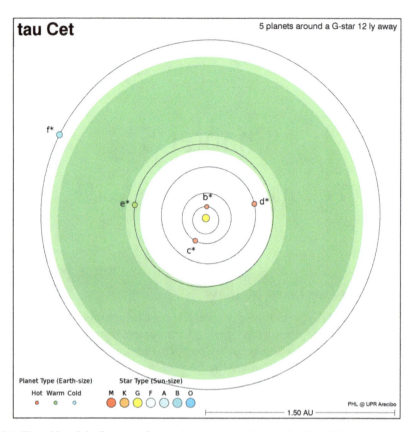

Figure 5.1. The orbits of the five unconfirmed planets around the star Tau Ceti. The green area shows the extent of the habitable zone. Planet 'e' is potentially habitable. Image courtesy of PHL @ UPR Arecibo.

Tau Ceti is a yellow star (a G-star like our Sun) with five still unconfirmed planets in the constellation Cetus at a distance of 12 light-years from Earth [1]. It has one potentially habitable planet discovered in 2012 (figure 5.1). The planet Tau Ceti e, with a minimum mass of four Earth masses, orbits close to the inner-edge of the habitable zone taking 168 days to orbit around its star. Tau Ceti e should be hotter than Earth assuming it has a similar terrestrial atmosphere. The star is visible to the naked eye (3.50 magnitude) and very popular in science fiction literature. This was one of the first stars monitored for signals of extraterrestrial intelligence at the beginning of the SETI project in 1960.

Kapteyn's Star is a red dwarf star (M-star) with two unconfirmed planets in the constellation Pictor at a distance of 13 light-years from Earth [2]. It has one potentially habitable planet discovered in 2014 (figure 5.2). The planet Kapteyn's Star b, with a minimum mass of five Earth masses, orbits inside the habitable zone in a very elliptical orbit taking 49 days to orbit around its star. Kapteyn's Star b should be colder than Earth assuming it has a similar terrestrial atmosphere. The star is not visible to the naked eye. Kapteyn's Star might be twice the age of our Sun. There has been recent debate about the existence of its planets [3].

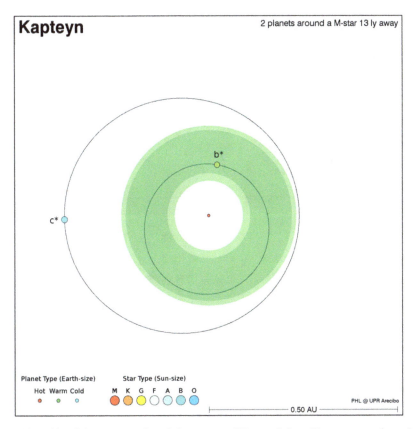

Figure 5.2. The orbits of the two unconfirmed planets around Kapteyn's Star. The green area shows the extent of the habitable zone. Planet 'b' is potentially habitable. Image courtesy of PHL @ UPR Arecibo.

HD 40307 is an orange star (K-star) with six planets, some still unconfirmed, in the constellation Pictor at a distance of 42 light-years from Earth [4]. It has one potentially habitable planet discovered in 2012 (figure 5.3). The planet HD 40307g, with a minimum mass of seven Earth masses, orbits inside the habitable zone in a very elliptical orbit taking 198 days to orbit around its star. HD 40307g should be colder than Earth assuming it has a similar terrestrial atmosphere. The star is not visible to the naked eye.

Gliese 3292 is a red dwarf star (M-star) with three confirmed planets in the constellation Eridanus at a distance of 59 light-years from Earth [5]. It has one potentially habitable planet discovered in 2014 (figure 5.4). The planet Gliese 3292c, with a minimum mass of nine Earth masses, orbits inside the habitable zone in a very elliptical orbit taking 48 days to orbit around its star. Gliese 3292c should be colder than Earth assuming it has a similar terrestrial atmosphere. The star is not visible to the naked eye.

Gliese 832 is a red dwarf star (M-star) with two confirmed planets in the constellation Grus at a distance of 16 light-years from Earth [6]. It has one potentially habitable planet discovered in 2014 (figure 5.5). The planet Gliese 832c, with a

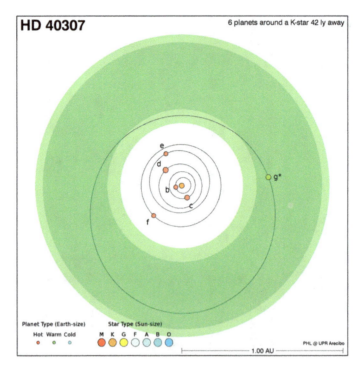

Figure 5.3. The orbits of the six planets around the star HD 40307. The green area shows the extent of the habitable zone. Planet 'g' is potentially habitable. Image courtesy of PHL @ UPR Arecibo.

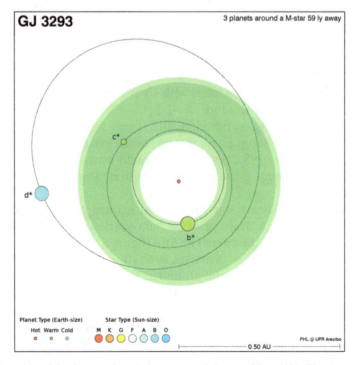

Figure 5.4. The orbits of the three confirmed planets around the star Gliese 3293. The green area shows the extent of the habitable zone. Planet 'c' is potentially habitable. Image courtesy of PHL @ UPR Arecibo.

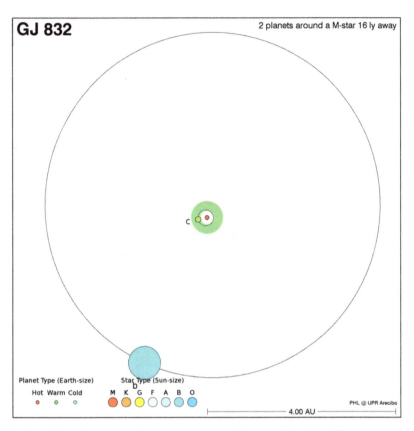

Figure 5.5. The orbits of the two confirmed planets around the star Gliese 832. The green area shows the extent of the habitable zone. Planet 'c' is potentially habitable. Image courtesy of PHL @ UPR Arecibo.

minimum mass of five Earth masses, orbits close to the inner-edge of the habitable zone in an elliptical orbit taking 36 days to orbit around its star. Gliese 832c should have temperatures similar to Earth assuming it has a similar terrestrial atmosphere. The star is not visible to the naked eye.

Gliese 682 is a red dwarf star (M-star) with two still unconfirmed planets in the constellation Scorpius at a distance of 17 light-years from Earth [7]. It has one potentially habitable planet discovered in 2014 (figure 5.6). The planet Gliese 682c, with a minimum mass of nine Earth masses, orbits inside the habitable zone in an elliptical orbit taking 57 days to orbit around its star. Gliese 682c should be colder than Earth assuming it has a similar terrestrial atmosphere. The star is not visible to the naked eye.

Gliese 667C is a red dwarf star (M-star) with up to six planets in the constellation Scorpius at a distance of 24 light-years from Earth [8]. It has up to three potentially habitable planet discovered in 2011 (figure 5.7). The planet Gliese 667Cc, with a minimum mass of four Earth masses, orbits inside the habitable zone taking 28 days to orbit around its star. Gliese 667Cc should have temperatures similar to Earth assuming it has a similar terrestrial atmosphere. Gliese 667Ce and Gliese 667Cf are

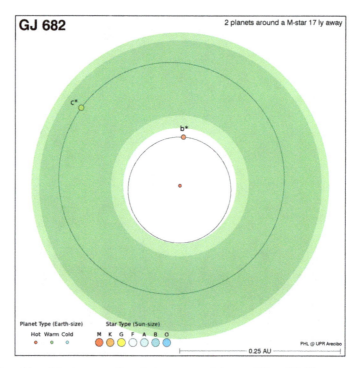

Figure 5.6. The orbits of the two unconfirmed planets around the star Gliese 682. The green area shows the extent of the habitable zone. Planet 'c' is potentially habitable. Image courtesy of PHL @ UPR Arecibo.

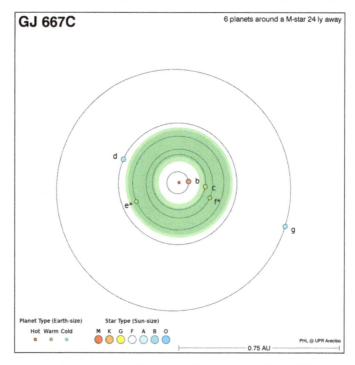

Figure 5.7. The orbits of the six planets (some still unconfirmed) around the star Gliese 667C. The green area shows the extent of the habitable zone. Planets 'c', 'e', and 'f' are potentially habitable. Image courtesy of PHL @ UPR Arecibo.

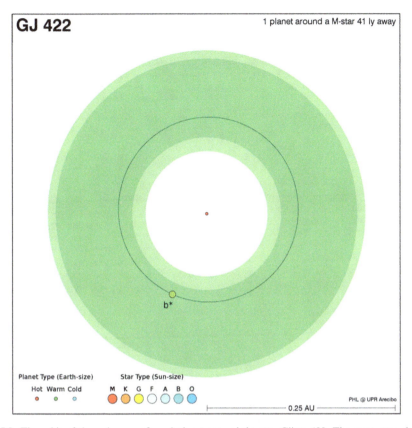

Figure 5.8. The orbit of the only unconfirmed planet around the star Gliese 422. The green area shows the extent of the habitable zone. Planet 'b' is potentially habitable. Image courtesy of PHL @ UPR Arecibo.

also potentially habitable but they are still unconfirmed. This planetary system is very interesting because it might have up to three habitable worlds around a nearby star. The star is not visible to the naked eye.

Gliese 422 is a red dwarf star (M-star) with one still unconfirmed planet in the constellation Centaurus at a distance of 41 light-years from Earth [9]. Its sole, potentially habitable planet was discovered in 2014 (figure 5.8). The planet Gliese 422b, with a minimum mass of ten Earth masses, orbits inside the habitable zone in a near-circular orbit taking 26 days to orbit around its star. Gliese 422b should have similar temperatures to Earth assuming it has a similar terrestrial atmosphere. The star is not visible to the naked eye.

Gliese 180 is a red dwarf star (M-star) with two still unconfirmed planets in the constellation Eridanus at a distance of 38 light-years from Earth [9]. Both planets are potentially habitable and were discovered in 2014 (figure 5.9). The planet Gliese 180b, with a minimum mass of eight Earth masses, orbits close to the inner-edge of the habitable zone in an elliptical orbit taking 17 days to orbit around its star. The planet Gliese 180c, with a minimum mass of six Earth masses, orbits inside the habitable zone taking 24 days. Gliese 180b and Gliese 180c should be hotter than

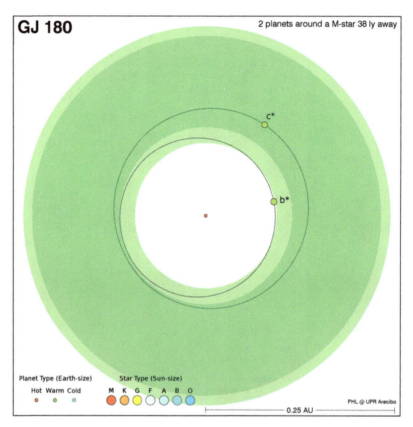

Figure 5.9. The orbits of the two unconfirmed planets around the star Gliese 180. The green area shows the extent of the habitable zone. Both planets 'b' and 'c' are potentially habitable. Image courtesy of PHL @ UPR Arecibo.

Earth assuming they have similar terrestrial atmospheres. The star is not visible to the naked eye.

Gliese 163 is a red dwarf star (M-star) with three confirmed planets in the constellation Dorado at a distance of 49 light-years from Earth [10]. It has one potentially habitable planet discovered in 2012 (figure 5.10). The planet Gliese 163c, with a minimum mass of seven Earth masses, orbits near the inner-edge of the habitable zone taking 26 days to orbit around its star. Gliese 163c should be hotter than Earth assuming it has a similar terrestrial atmosphere. The star is not visible to the naked eye.

5.2 The planets we know by size

The transit method is used to measure the size of planets given that they pass in front of their star. The following paragraphs describe all known stars with planets that might be considered potentially habitable because they have a radius less than 2.5 Earth radii and orbit within the habitable zone. Thus liquid water, if available, might be stable at their surfaces. Planets larger than 2.5 Earth radii are expected to be non-habitable gas worlds but there is no guarantee that those below this value

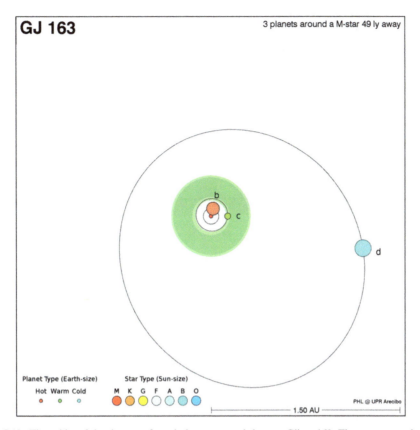

Figure 5.10. The orbits of the three confirmed planets around the star Gliese 163. The green area shows the extent of the habitable zone. Planet 'c' is potentially habitable. Image courtesy of PHL @ UPR Arecibo.

are indeed habitable rocky worlds. We do not know if they are rocky, water, or gas worlds since we do not know their mass to estimate their bulk density. Just recently astronomers were able to also measure the mass of one of these planets (figure 5.25).

Kepler-452 is a yellow star (a G-star like our Sun) with only one confirmed planet in the constellation Cygnus at a distance of 1402 light-years from Earth [11]. Its sole, potentially habitable planet was discovered in 2015 (figure 5.11). The planet Kepler-452b, with a radius of 1.6 Earth radii, orbits inside the habitable zone taking 385 days to orbit around its star. Kepler-452b should be hotter than Earth assuming it has a similar terrestrial atmosphere. The star is not visible to the naked eye.

Kepler-443 is an orange star (K-star) with only one confirmed planet in the constellation Cygnus at a distance of 2540 light-years from Earth [12]. Its sole, potentially habitable planet was discovered in 2015 (figure 5.12). The planet Kepler-443b, with a radius of 2.3 Earth radii, orbits inside the habitable zone taking 178 days to orbit around its star. Kepler-443b should be slightly colder than Earth assuming it has a similar terrestrial atmosphere. The star is not visible to the naked eye.

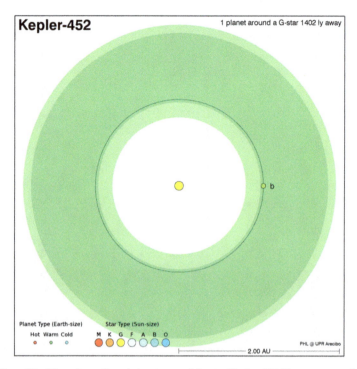

Figure 5.11. The orbit of the only confirmed planet around the star Kepler-452. The green area shows the extent of the habitable zone. Its only planet 'b' is potentially habitable. Image courtesy of PHL @ UPR Arecibo.

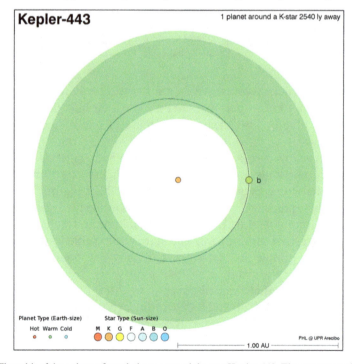

Figure 5.12. The orbit of the only confirmed planet around the star Kepler-443. The green area shows the extent of the habitable zone. Its only planet 'b' is potentially habitable. Image courtesy of PHL @ UPR Arecibo.

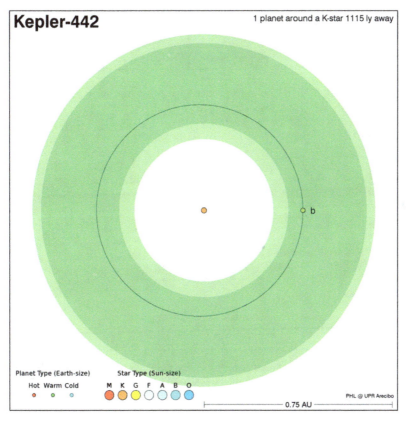

Figure 5.13. The orbit of the only confirmed planet around the star Kepler-442. The green area shows the extent of the habitable zone. Its only planet 'b' is potentially habitable. Image courtesy of PHL @ UPR Arecibo.

Kepler-442 is an orange star (K-star) with only one confirmed planet in the constellation Lyra at a distance of 1115 light-years from Earth [12]. Its sole, potentially habitable planet was discovered in 2015 (figure 5.13). The planet Kepler-442b, with a radius of 1.3 Earth radii, orbits inside the habitable zone taking 112 days to orbit around its star. Kepler-442b should be colder than Earth assuming it has a similar terrestrial atmosphere. The star is not visible to the naked eye.

Kepler-440 is an orange star (K-star) with only one confirmed planet in the constellation Lyra at a distance of 1115 light-years from Earth [12]. Its sole, potentially habitable planet was discovered in 2015 (figure 5.14). The planet Kepler-440b, with a radius of 1.9 Earth radii, orbits inside the habitable zone taking 101 days to orbit around its star. Kepler-440b should be hotter than Earth assuming it has a similar terrestrial atmosphere. The star is not visible to the naked eye.

Kepler-438 is an orange star (K-star) with only one confirmed planet in the constellation Lyra at a distance of 473 light-years from Earth [12]. Its sole, potentially habitable planet was discovered in 2015 (figure 5.15). The planet

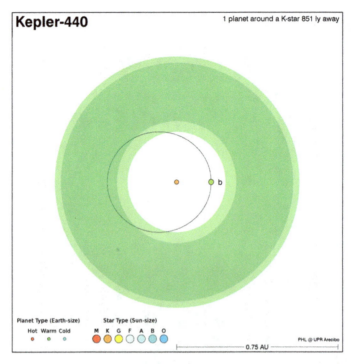

Figure 5.14. The orbit of the only confirmed planet around the star Kepler-440. The green area shows the extent of the habitable zone. Its only planet 'b' is potentially habitable. Image courtesy of PHL @ UPR Arecibo.

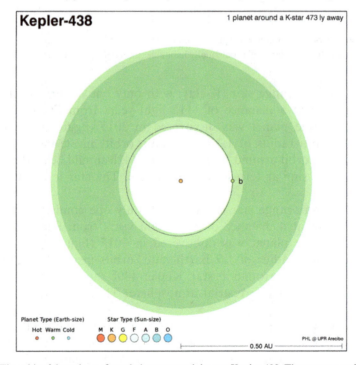

Figure 5.15. The orbit of the only confirmed planet around the star Kepler-438. The green area shows the extent of the habitable zone. Its only planet 'b' is potentially habitable. Image courtesy of PHL @ UPR Arecibo.

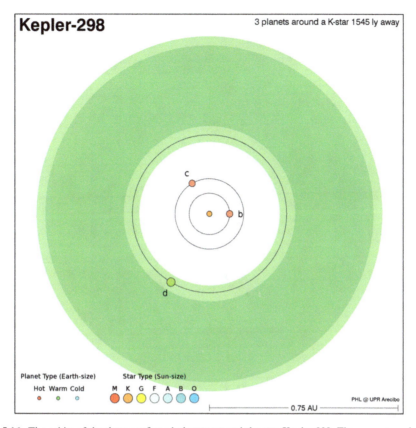

Figure 5.16. The orbits of the three confirmed planets around the star Kepler-298. The green area shows the extent of the habitable zone. Its planet 'd' is potentially habitable. Image courtesy of PHL @ UPR Arecibo.

Kepler-438b, with a radius of 1.1 Earth radii, orbits close to the inner-edge of the habitable zone taking 35 days to orbit around its star. Kepler-438b should be hotter than Earth assuming it has a similar terrestrial atmosphere. The star is not visible to the naked eye.

Kepler-298 is an orange star (K-star) with three confirmed planets in the constellation Lyra at a distance of 1545 light-years from Earth [13]. It has one potentially habitable planet discovered in 2014 (figure 5.16). The planet Kepler-298d, with a radius of 2.5 Earth radii, orbits close to the inner-edge of the habitable zone taking 78 days to orbit around its star. Kepler-298d should be hotter than Earth assuming it has a similar terrestrial atmosphere. The star is not visible to the naked eye.

Kepler-296 is an orange star (K-star) with five confirmed planets in the constellation Lyra at a distance of 737 light-years from Earth [13]. It has one potentially habitable planet discovered in 2015 (figure 5.17). The planet Kepler-296f, with a radius of 1.8 Earth radii, orbits close to the inner-edge of the habitable zone taking 63 days to orbit around its star. Kepler-296f should be

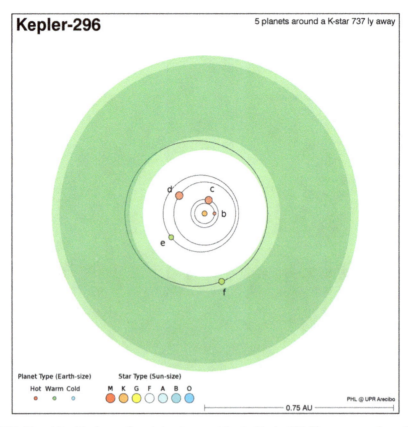

Kepler-296

5 planets around a K-star 737 ly away

Planet Type (Earth-size) Star Type (Sun-size)

Hot Warm Cold   M K G F A B O

PHL @ UPR Arecibo

0.75 AU

Figure 5.17. The orbits of the five confirmed planets around the star Kepler-296. The green area shows the extent of the habitable zone. Its planet 'f' is potentially habitable. Image courtesy of PHL @ UPR Arecibo.

hotter than Earth assuming it has a similar terrestrial atmosphere. The star is not visible to the naked eye.

Kepler-283 is an orange star (K-star) with two confirmed planets in the constellation Cygnus at a distance of 1741 light-years from Earth [13]. It has one potentially habitable planet discovered in 2014 (figure 5.18). The planet Kepler-283c, with a radius of 1.8 Earth radii, orbits close to the inner-edge of the habitable zone taking 93 days to orbit around its star. Kepler-283c should have similar temperatures to Earth assuming it has a similar terrestrial atmosphere. The star is not visible to the naked eye.

Kepler-186 is an orange star (K-star) with five confirmed planets in the constellation Cygnus at a distance of 561 light-years from Earth [13]. It has one potentially habitable planet discovered in 2015 (figure 5.19). The planet Kepler-186f, with a radius of 1.8 Earth radii, orbits close to the outer-edge of the habitable zone taking 130 days to orbit around its star. Kepler-186f should be much colder than Earth assuming it has a similar terrestrial atmosphere. The star is not visible to the naked eye.

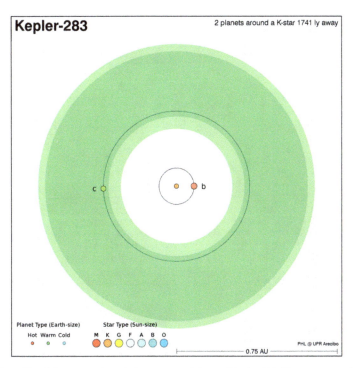

Figure 5.18. The orbits of the two confirmed planets around the star Kepler-283. The green area shows the extent of the habitable zone. Its planet 'c' is potentially habitable. Image courtesy of PHL @ UPR Arecibo.

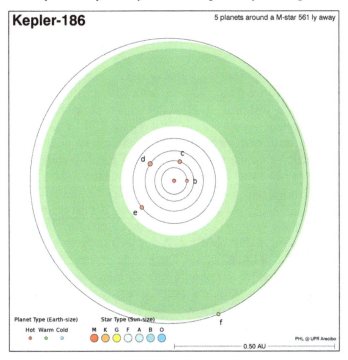

Figure 5.19. The orbits of the five confirmed planets around the star Kepler-186. The green area shows the extent of the habitable zone. Its planet 'f' is potentially habitable. Image courtesy of PHL @ UPR Arecibo.

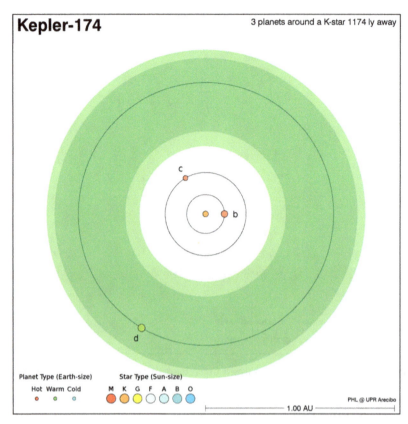

Figure 5.20. The orbits of the three confirmed planets around the star Kepler-174. The green area shows the extent of the habitable zone. Its planet 'd' is potentially habitable. Image courtesy of PHL @ UPR Arecibo.

Kepler-174 is an orange star (K-star) with three confirmed planets in the constellation Lyra at a distance of 1174 light-years from Earth [13]. It has one potentially habitable planet discovered in 2014 (figure 5.20). The planet Kepler-174d, with a radius of 2.2 Earth radii, orbits in the habitable zone taking 247 days to orbit around its star. Kepler-174d should be colder than Earth assuming it has a similar terrestrial atmosphere. The star is not visible to the naked eye.

Kepler-62 is an orange star (K-star) with five confirmed planets in the constellation Lyra at a distance of 1200 light-years from Earth [14]. It has two potentially habitable planets discovered in 2013 (figure 5.21). The planet Kepler-62e, with a radius of 1.6 Earth radii, orbits in the habitable zone taking 122 days to orbit around its star. The planet Kepler-62f has radius of 1.4 Earth radii and orbits inside the habitable zone taking 267 days. Kepler-62e should have similar temperatures to Earth and Kepler-62f should be colder assuming both have a similar terrestrial atmosphere. The star is not visible to the naked eye.

Kepler-61 is an orange star (K-star) with only one confirmed planet in the constellation Cygnus at a distance of 1063 light-years from Earth [15]. Its sole potentially habitable planet was discovered in 2012 (figure 5.22). The planet

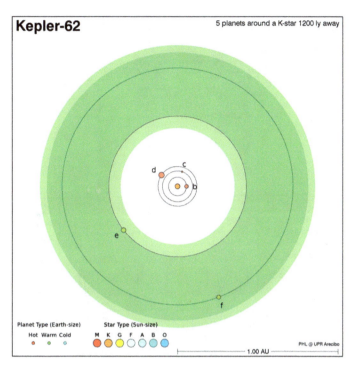

Figure 5.21. The orbits of the five confirmed planets around the star Kepler-62. The green area shows the extent of the habitable zone. Planets 'e' and 'f' are potentially habitable. Image courtesy of PHL @ UPR Arecibo.

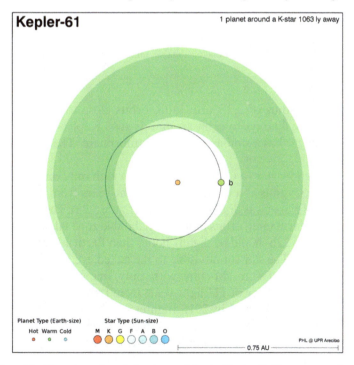

Figure 5.22. The orbit of the only confirmed planet around the star Kepler-61. The green area shows the extent of the habitable zone. Its only planet 'b' is potentially habitable. Image courtesy of PHL @ UPR Arecibo.

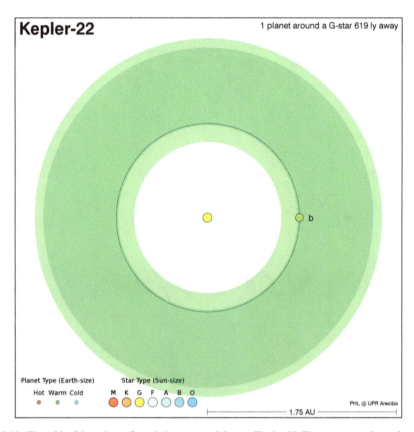

Figure 5.23. The orbit of the only confirmed planet around the star Kepler-22. The green area shows the extent of the habitable zone. Its only planet 'b' is potentially habitable. Image courtesy of PHL @ UPR Arecibo.

Kepler-61b, with a radius of 2.2 Earth radii, orbits close to the inner-edge of the habitable zone taking 60 days to orbit around its star. Kepler-61b should be hotter than Earth assuming it has a similar terrestrial atmosphere. The star is not visible to the naked eye.

Kepler-22 is a yellow star (a G-star like our Sun) with only one confirmed planet in the constellation Cygnus at a distance of 619 light-years from Earth [16]. Its sole, potentially habitable planet was discovered in 2011 (figure 5.23). The planet Kepler-22b, with a radius of 2.4 Earth radii, orbits inside the habitable zone taking 290 days to orbit around its star. Kepler-22b should be slightly hotter than Earth assuming it has a similar terrestrial atmosphere. The star is not visible to the naked eye.

K2-18 is a red dwarf star (M-star) with only one confirmed planet in the constellation Leo at a distance of 111 light-years from Earth [17]. Its sole, potentially habitable planet was discovered in 2015 (figure 5.24). The planet K2-18b, with a radius of 2.2 Earth radii, orbits inside the habitable zone taking 33 days to orbit around its star. K2-18b should have similar temperatures to Earth assuming it has a similar terrestrial atmosphere. The star is not visible to the naked eye.

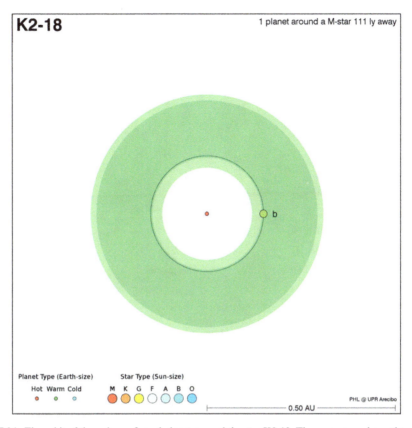

Figure 5.24. The orbit of the only confirmed planet around the star K2-18. The green area shows the extent of the habitable zone. Its only planet 'b' is potentially habitable. Image courtesy of PHL @ UPR Arecibo.

K2-3 is a red dwarf star (M-star) with three confirmed planets in the constellation Leo at a distance of 147 light-years from Earth [18]. It has one potentially habitable planet discovered in 2015 (figure 5.25). The planet K2-3d, with a radius of 1.5 Earth radii, orbits close to the inner-edge of the habitable zone taking 45 days to orbit around its star. K2-3d should be much hotter than Earth assuming it has a similar terrestrial atmosphere. The star is not visible to the naked eye. New observations with HARPS also show that this planet has a mass of eleven Earth masses thus making it a rocky planet much denser than Earth [19]. This is the first planet in the habitable zone that is known by both mass and radius.

Wolf 1061 is a red dwarf star (M-star) with three confirmed planets in the constellation Ophiuchus at a distance of 14 light-years from Earth [20]. It has one potentially habitable planet discovered in 2015 (figure 5.26). The planet K2-3d, with a radius of 1.5 Earth radii, orbits close to the inner-edge of the habitable zone taking 45 days to orbit around its star. K2-3d should be much hotter than Earth assuming it has a similar terrestrial atmosphere. The star is not visible to the naked eye. Wolf 1061 is also the closest system with confirmed potentially habitable planets. Those of Tau Ceti and Kapteyn's Star are closer but are still unconfirmed.

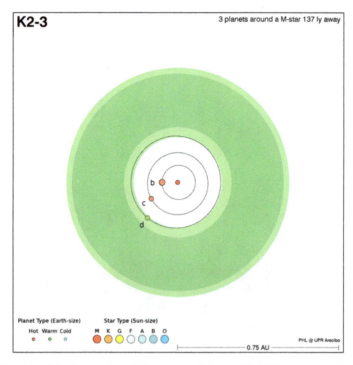

Figure 5.25. The orbits of the three confirmed planets around the star K2-3. The green area shows the extent of the habitable zone. Its planet 'd' is potentially habitable. Image courtesy of PHL @ UPR Arecibo.

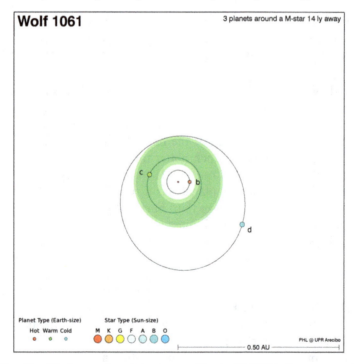

Figure 5.26. The orbits of the three confirmed planets around the star Wolf 1061. The green area shows the extent of the habitable zone. Its planet 'c' is potentially habitable. Image courtesy of PHL @ UPR Arecibo.

References

[1] Tuomi M *et al* 2013 Signals embedded in the radial velocity noise: periodic variations in the τ Ceti velocities★ *Astron. Astrophys.* **551** A79

[2] Anglada-Escude G *et al* 2014 Two planets around Kapteyn's star: a cold and a temperate super-Earth orbiting the nearest halo red dwarf *Mon. Not. R. Astron. Soc.: Lett.* **443**(1) L89–L93

[3] Robertson P, Roy A and Mahadevan S 2015 Stellar activity mimics a habitable-zone planet around Kapteyn's Star *Astrophys. J. Lett.* **805**(2) L22

[4] Tuomi M *et al* 2013 Habitable-zone super-Earth candidate in a six-planet system around the K2.5V star HD 40307 *Astron. Astrophys.* **549** A48

[5] Astudillo-Defru N *et al* 2015 The HARPS search for southern extra-solar planets: XXXVI. Planetary systems and stellar activity of the M dwarfs GJ 3293, GJ 3341, and GJ 3543★★★ *Astron. Astrophys.* **575** A119

[6] Wittenmyer R A *et al* 2014 GJ 832c: a super-earth in the habitable zone *Astrophys. J.* **791**(2) 114

[7] Tuomi M, Jones H R A, Barnes J R, Anglada-Escudé G and Jenkins J S 2014 Bayesian search for low-mass planets around nearby M dwarfs—estimates for occurrence rate based on global detectability statistics *Mon. Not. R. Astron. Soc.* **441**(2) 1545–69

[8] Anglada-Escudé G *et al* 2013 A dynamically packed planetary system around GJ 667C with three super-Earths in its habitable zone *Astron. Astrophys.* **556** A126

[9] Tuomi M, Jones H R A, Barnes J R, Anglada-Escudé G and Jenkins J S 2014 Bayesian search for low-mass planets around nearby M dwarfs—estimates for occurrence rate based on global detectability statistics *Mon. Not. R. Astron. Soc.* **441**(2) 1545–69

[10] Bonfils X *et al* 2013 The HARPS search for southern extra-solar planets: XXXIV. A planetary system around the nearby M dwarf GJ 163, with a super-Earth possibly in the habitable zone *Astron. Astrophys.* **556** A110

[11] Jenkins J M *et al* 2015 Discovery and validation of Kepler-452b: a 1.6 $R_\oplus$ super Earth exoplanet in the habitable zone of a G2 star *Astron. J.* **150**(2) 56

[12] Torres G *et al* 2015 Validation of twelve small Kepler transiting planets in the habitable zone *Astrophys. J.* **800**(2) 99

[13] Rowe J F *et al* 2014 Validation of Kepler's multiple planet candidates: III. Light curve analysis and announcement of hundreds of new multi-planet systems *Astrophys. J.* **784**(1) 45

[14] Borucki W J *et al* 2013 Kepler-62: a five-planet system with planets of 1.4 and 1.6 Earth radii in the habitable zone *Science* **340**(6132) 587–90

[15] Ballard S *et al* 2013 Exoplanet characterization by proxy: a transiting 2.15 $R_\oplus$ planet near the habitable zone of the late K dwarf Kepler-61 *Astrophys. J.* **773**(2) 98

[16] Borucki W J *et al* 2012 Kepler-22b: a 2.4 Earth-radius planet in the habitable zone of a Sun-like star *Astrophys. J.* **745**(2) 120

[17] Montet B T *et al* 2015 Stellar and planetary properties of K2 Campaign 1 candidates and validation of 17 planets, including a planet receiving Earth-like insolation *Astrophys. J.* **809**(1) 25

[18] Crossfield I J M *et al* 2015 A nearby M star with three transiting super-Earths discovered by K2 *Astrophys. J.* **804**(1) 10

[19] Almenara J M *et al* 2015 A HARPS view on K2-3 *Astron. Astrophys.* **581** L7

[20] Wright D J, Wittenmyer R A, Tinney C G, Bentley J S and Zhao J 2015 Three planets orbiting Wolf 1061, arXiv: 1512.05154

Chapter 6

And the search goes on...

6.1 The observatories

Today we are detecting and characterizing potentially habitable worlds at an increasing rate. In the next decades we might reach a stopping point where our technology to observe these worlds may stall, still providing better details for many centuries to come. Our ability to explore these worlds is limited by observations from afar and traveling to these worlds is only possible in sci-fi adventures. It might take centuries before we develop the technology to travel to these worlds, even using unmanned robotic explorers.

Many observatories around the world are used to search for exoplanets. Most are ground telescopes because space telescopes are much more difficult and expensive to build. Not all have the sensitivity to detect potentially habitable planets, i.e. Earth-size planets in the habitable zone. So far only two observatories have been able to detect these interesting worlds: ESO's HARPS and NASA's Kepler Space Telescope.

The following sections describe the most notable space telescopes that will be detecting and studying potentially habitable worlds in the next decade [1]. Kepler, in its K2 extended mission, is currently in service and TESS (Transiting Exoplanet Survey Satellite), JWST (James Webb Space Telescope) and Plato will soon join in the exploration. There are other planned space telescopes such as the WFIRST-AFTA (Wide-Field Infrared Survey Telescope-Astrophysics Focused Telescope Assets) and the New Worlds Telescopes but these are still under early development. Present and future ground telescopes such as the European Extremely Large Telescope will also complement the observations of the space telescopes.

6.2 NASA's Kepler

The NASA Kepler Mission is a space telescope that uses the transit method to survey a portion of the night sky to discover planets, in particular Earth-size planets in or near the habitable zone (figure 6.1). It completed its primary mission from 2009–12

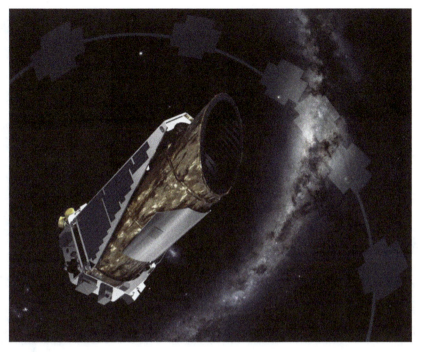

Figure 6.1. Artistic representation of the NASA Kepler space telescope and the regions of space that it is currently observing as part of its extended K2 mission. Image courtesy of NASA.

and it is now operating in an extended mode under the name of the K2 Mission. Results from Kepler are used to place our Solar System within the context of planetary systems in our Galaxy. Kepler looks at more than 145 000 stars so if Earth-size planets are common, then Kepler should detect hundreds of them.

Scientists want to find planets in the habitable zone of stars like the Sun. This means that the time between transits is about one year. To reliably detect a sequence one needs four transits. Hence, the mission duration needs to be at least three and a half years. If the Kepler Mission continues for longer, it will be able to detect smaller, and more distant planets as well as a larger number of true Earth analogs. The telescope must be in space to obtain the photometric precision needed to reliably see an Earth-like transit and to avoid interruptions caused by day–night cycles, seasonal cycles, and atmospheric perturbations.

Kepler is a specially designed 0.95 m diameter telescope called a photometer or light meter. It has a very large field of view for an astronomical telescope, 105 square degrees, which is comparable to the area of your hand held at arm's length. The field of view of most telescopes is less than one square degree. Kepler has a large field of view in order to observe more than 145 000 stars simultaneously to monitor their brightness for at least 3.5 years, the initial length of the mission.

Kepler and its K2 follow-up observations found over 1000 confirmed exoplanets in over 400 stellar systems, along with over 3000 still unconfirmed planet candidates. Up to 16 of these confirmed planets might be potentially habitable (listed in chapter 5)

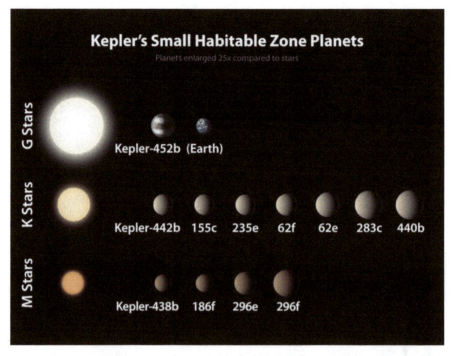

Figure 6.2. These are the potentially habitable worlds discovered by NASA Kepler that are more similar to Earth in size and insolation. Only Kepler-452b orbits a similar G star like our Sun. Image courtesy of NASA.

but over one hundred are still waiting confirmation. One of the most interesting is Kepler-452b (figure 6.2).

6.3 NASA's TESS

The Transiting Exoplanet Survey Satellite (TESS) is a planned space telescope for 2017. It will use the transit method to detect exoplanets (figure 6.3). It is the first-ever space telescope capable of an all-sky transit survey (Kepler is limited to smaller regions of the sky). TESS will identify planets ranging from Earth-sized to gas giants, orbiting a wide range of stellar types and orbital distances. The principal goal of the TESS mission is to detect small planets with bright host stars in the solar neighborhood, so that detailed characterizations of the planets and their atmospheres can be performed by other telescopes.

TESS will monitor the brightness of more than 500 000 stars during its two-year mission, searching for temporary drops in brightness caused by planetary transits. It is expected to catalog more than three thousand transiting exoplanet candidates, including a sample of about 500 Earth-sized and 'super-Earth' planets, with radii less than twice that of the Earth. TESS will be able to detect small rock-and-ice planets orbiting a diverse range of stellar types and covering a wide span of orbital periods, including rocky worlds in the habitable zones of their host stars.

TESS stars will be 30 to 100 times brighter than those surveyed by Kepler, which are far easier to characterize with follow-up observations. These observations will

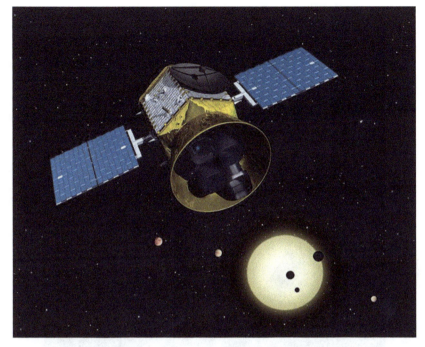

Figure 6.3. Artistic representation of the planned TESS space telescope. Image courtesy of NASA.

provide refined measurements of the planet masses, sizes, densities and atmospheric properties. TESS will provide prime targets for further, more detailed characterization with the James Webb Space Telescope (JWST), as well as other large ground-based and space-based telescopes of the future. TESS's legacy will be a catalog of the nearest and brightest stars hosting transiting exoplanets, which will comprise the most favorable targets for detailed investigation in the coming decades.

6.4 ESA's Plato

The Planetary Transits and Oscillations of stars (PLATO) is a space telescope planned for 2024 that will use the transit method to detect exoplanets (figure 6.4). The primary goal of PLATO is to detect exoplanets in the habitable zone of solar-type stars and characterize their bulk properties. PLATO will provide the key information (planet radii, mean densities, stellar irradiation and architecture of planetary systems) needed to determine the habitability of these expectedly diverse new worlds.

PLATO will characterize thousands of rocky (including Earth twins), icy or giant planets by providing precise measurements of their radii, masses and ages. In total, the PLATO catalogue will consist of thousands of characterized planets of all types and over 85 000 stars with accurately known ages and masses. Planets discovered around the bright PLATO stars will be prime targets for spectroscopic transit follow-up observations of their atmospheres.

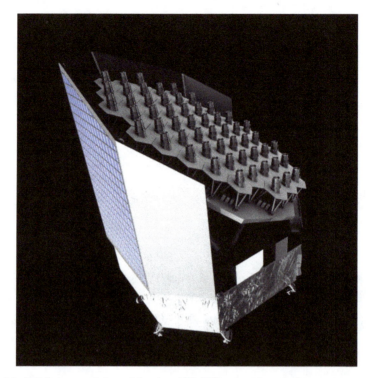

Figure 6.4. Artistic representation of the planed PLATO space telescope. Image courtesy of ESA.

The PLATO catalog will therefore play a key role in identifying small planet targets of interest at intermediate orbital distances. It will also provide information on planetary albedos and the stratification of planetary atmospheres. Finally, the close-in planets found around stars of different types and ages will provide a huge sample to study the interaction between stars and planets. PLATO is very similar to TESS but has a longer primary mission.

6.5 NASA JWST

The JWST is a space telescope under construction and scheduled to launch in October 2018 (figure 6.5). One of its main objectives is to study the atmospheres of exoplanets and search for the building blocks of life elsewhere in the Universe. One method JWST will use for studying exoplanets is the transit method to perform spectroscopy of the atmosphere of planets. JWST will also carry coronagraphs to enable the direct imaging of exoplanets near bright stars. The image of the exoplanet would just be a dot, but enough to observe color variations due to seasons, rotation and weather.

The JWST is a successor of the Hubble Space Telescope and the Spitzer Space Telescope and will offer unprecedented resolution and sensitivity from visible–infrared light. The telescope has a segmented 6.5 m diameter primary mirror and a large sunshield keeping its mirror and four science instruments at very low temperatures to improve their sensitivity.

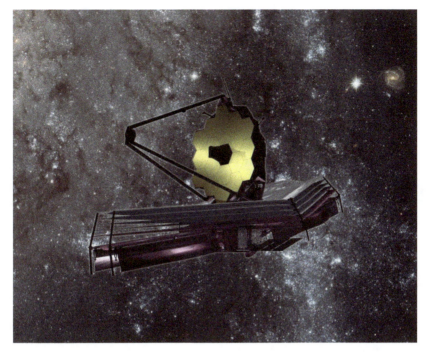

Figure 6.5. Artistic representation of the JWST. Image courtesy of NASA.

JWST is a multi-purpose space telescope, just like Hubble, that will enable a broad range of investigations across the fields of astronomy and cosmology. One of its goals is to understand the formation of stars and planets. This will include imaging molecular clouds and star-forming clusters, studying the debris disks around stars, direct imaging of exoplanets and spectroscopic examination of planetary transits.

Reference

[1] Boss A P, Hudgins D M and Traub W A 2010 New worlds, new horizons and NASA's approach to the next decade of exoplanet discoveries *The Astrophysics of Planetary Systems: Formation, Structure, and Dynamical Evolution* **6** 324–34

CPSIA information can be obtained
at www.ICGtesting.com
Printed in the USA
BVHW020751210323
660845BV00003B/208